THE
STRATEGIC
PROJECT
PLANNER

THE STRATEGIC PROJECT PLANNER

A Profit-Driven Project Management℠ Process for Planning Projects to Meet Business Goals

Richard E. Westney

Westney Project Services, Inc.
Houston, Texas

MARCEL DEKKER, INC.   NEW YORK • BASEL

ISBN: 0-8247-0392-8

This book is printed on acid-free paper.

Headquarters
Marcel Dekker, Inc.
270 Madison Avenue, New York, NY 10016
tel: 212-696-9000; fax: 212-685-4540

Eastern Hemisphere Distribution
Marcel Dekker AG
Hutgasse 4, Postfach 812, CH-4001 Basel, Switzerland
tel: 41-61-261-8482; fax: 41-61-261-8896

World Wide Web
http://www.dekker.com

The publisher offers discounts on this book when ordered in bulk quantities. For more information, write to Special Sales/Professional Marketing at the headquarters address above.

Current printing (last digit):
10 9 8 7 6 5 4 3 2 1

PRINTED IN THE UNITED STATES OF AMERICA

This book is dedicated to each of the fine people of the Westney Project Services team. No project manager could ask for more.

About the Author

Richard E. Westney is known worldwide as an authority on project management. Author of four books on the subject, he has taught project management at major universities such as Texas A&M, Stanford, and the Norwegian Institute of Technology. Over 12,000 people from 65 countries have completed his courses on applied project management.

Mr. Westney is founder and CEO of Westney Project Services, Inc., in Houston, Texas, a company he formed in 1978 after participating in worldwide energy projects for Exxon. He holds a BS in mechanical engineering from the City University of New York, and an MS in management science from Rensselaer Polytechnic Institute. He is Past President and Fellow of the Association for the Advancement of Cost Engineering (AACE International), and an active member of the Project Management Institute (PMI).

Mr. Westney is a licensed professional engineer in Texas and New Jersey, and a certified Project Management Professional.

Preface

The idea of Strategic Project Planning is really nothing new. What *is* new is the recognition of the immense value that the time and effort spent in Strategic Project Planning have for the project team and the people who must live with the result.

For the past 10 years or so, project teams have been gathering to spend from one to several days going through the Strategic Project Planning Process to develop a Project Execution Plan. As facilitators for this process, the people of Westney Project Services have developed and continuously improved methods to make this process both effective and efficient.

Strategic Project Planning is certainly not easy. Many project team members are uncomfortable with the ambiguity associated with thinking about strategies early in a project's development. We wish we could just get down to business—and jump to the "execution" phase of the project as soon as possible. But today's fast-paced, business-focused projects demand that we get the planning done right. The best project managers know that *the shorter the schedule, the more important it is to take the time to plan.*

But exactly how should this be done? The Strategic Project Planner uses a proven question-and-answer method to focus the team's thinking on the most important elements and issues. This Q&A method was developed over a period of 10 years, based on our experience in facilitating planning workshops with hundreds of projects of all types. Over and over, we have seen these workshops help project teams to:

- Make decisions that saved millions of dollars
- Resolve major conflicts and find a way for team members to work together
- Change the strategic direction of major projects
- Identify and find a way to manage risks

We started using the first version of the Strategic Project Planner in 1994, and it has become a standard planning tool for many projects and organizations. This new edition incorporates everything we have learned from the many project teams with whom we have been privileged to work. They have seen that this method works, and we know it will become one of your favorite project management tools too.

Richard E. Westney, P.E.

Contents

I. How Strategic Project Planning Ensures Success

A. Benefits of Strategic Project Planning

Leading organizations have repeatedly proven the bottom-line benefits of the effective application of modern project management methods. Projects are the means by which an organization stays competitive: by introducing new products, increasing productivity, and gaining knowledge.

An organization's capability and effectiveness at planning and executing projects have a direct impact on profitability. A recent study by the Business Roundtable showed that organizations that played a strong leadership role in managing their projects, and that applied Strategic Project Planning and related Best Practices, enjoyed significantly lower project costs, shorter schedules, higher quality, and increased profitability (see Figure 1).

It is interesting to note that the conventional wisdom—that no project can be achieved with low cost, a fast schedule, and high quality all at the same

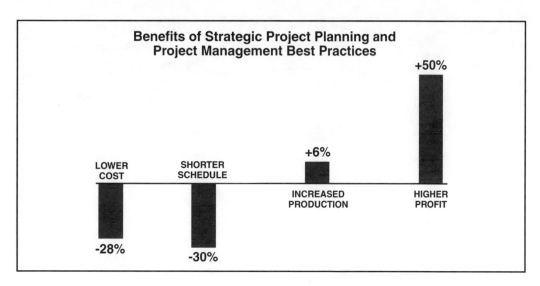

Figure 1: Benefits of Strategic Project Planning

time—has been overturned. Through the use of Strategic Project Planning, and the effective implementation of Project Management Best Practices, today's project managers carry out projects that excel in all three categories.

B. What Is Strategic Project Planning?

We all perform projects, and we all want our projects to succeed. What is Strategic Project Planning, and how will it help us accomplish this goal?

1. What is a project?

A good way to explain Strategic Project Planning is to start by defining what we mean by a project:

> *A project is the work that transforms an __opportunity__ into an __asset__.*

For example:

Opportunity	Project	Asset
Sell a new product	Design and build a manufacturing facility	Manufacturing facility producing the required product
Improve office productivity	Design and implement a Management Information System	Hardware and software working effectively in an operational information system
Acquire new knowledge	Design and execute a research & development program	New knowledge that provides competitive advantage in product development
Increase employee job satisfaction	Design and implement a new benefits package	An effective benefit plan

2. What is "project success"?

Traditional project managers might consider a successful project to be one that is completed on time and on (or below) budget. Although these may be characteristics of a successful project, a project that meets these objectives is not necessarily a success.

Today's project managers know that the only way to define project success is as follows:

> *A successful project is one that meets its <u>business goals</u>.*

A project may finish on time and on budget, but it will not be successful unless the business goals are met. Project managers must therefore focus at all times on the business goals that gave rise to the project in the first place.

3. What is Strategic Project Planning?

Strategic Project Planning is the first step toward attaining the benefits shown in Figure 1. We can define it as follows:

> *Strategic Project Planning is the process of defining how a project will be executed in order to achieve its business goals.*

The objective of Strategic Project Planning is to produce a *Project Execution Plan*, which is used to guide plans, decisions and actions throughout the project's life cycle.

Strategic Project Planning allows us to take advantage of the critical early stages of the "Influence Curve." The Influence Curve, shown in Figure 2, demonstrates that the most important time for ensuring a project's success is in its early phases. Yet this is the part of a project that is most often neglected.

What should we be doing in the early phases of a project in order to exert the maximum influence on project success? The answer is defined through Strategic Project Planning. The resulting Project Execution Plan will show how the project will be organized and led in order to implement Best Practices, and meet its business goals.

4. What is a Project Execution Plan?

A Project Execution Plan (PEP) is a document that describes the project team's strategy for planning and executing the project in order to meet its business goals. It ensures alignment at all times and at all levels, between project plans, decisions and actions, and the business goals that led to the project in the first place.

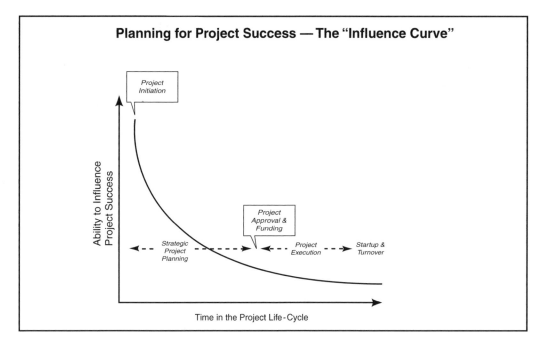

*Figure 2: The Influence Curve showing the importance of
early Strategic Project Planning in ensuring project success*

The PEP is a living document, periodically updated to reflect changes in
business conditions, priorities, or strategies. It is a guide for everyone in the
project, be they members of the project team, investors or partners, or
contractors serving the project. It defines roles and responsibilities, identifies issues, and serves as a tool for communication.

The PEP is best presented as a summary document, incorporating detailed
information, such as relevant procedures, by reference. It gives a "big pic-
ture" view of what the project is, why it is important, and the strategy for
executing it.

II. The Strategic Project Planning Process

Strategic Project Planning is a three-phase process (see Figure 3) that evolves over the early stages of a project and is completed with the preparation of a fully developed Project Execution Plan at project funding.

A project usually begins with the recognition of an opportunity and the development of the goals to be achieved. This is also the first step in the Strategic Project Planning Process: Defining the Vision of Success.

Defining the *Vision of Success* means defining the business goals that will determine project success, and translating these into project objectives and a scope of work.

Once the Vision of Success has been developed, we can begin defining the *Strategy for Success*. This requires defining the major phases and milestones, the risks to be managed, how the project will be organized, how contractors will be employed, what Best Practices are to be used, and how the project team will achieve high performance.

The Vision of Success and the Strategy for Success provide the basis for Defining the *Tools for Success*. In this section of the Project Execution Plan, we will define how time, cost, quality, resources, and safety will be managed.

We can expand the phases into a complete, 15-step Strategic Project Planning Process as shown in Figure 4.

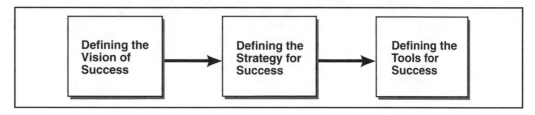

Figure 3: The three-phase Strategic Project Planning Process

II. The Strategic Project Planning Process

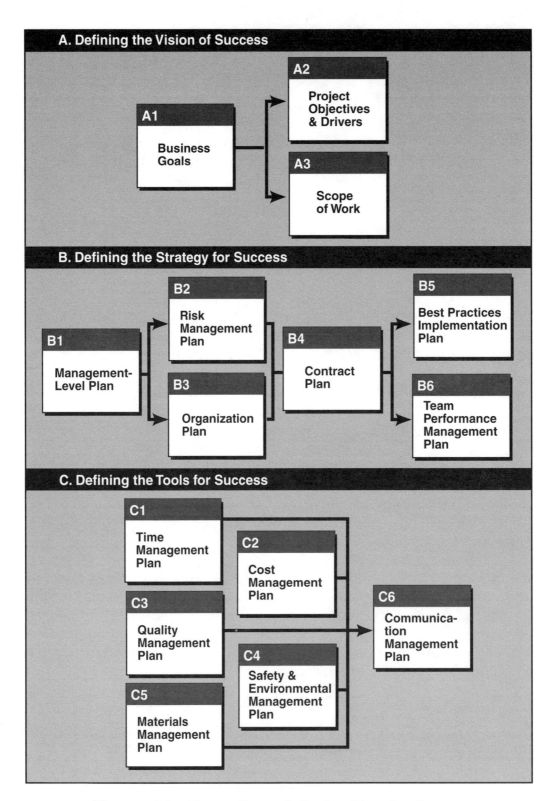

Figure 4: The 15-step Strategic Project Planning Process

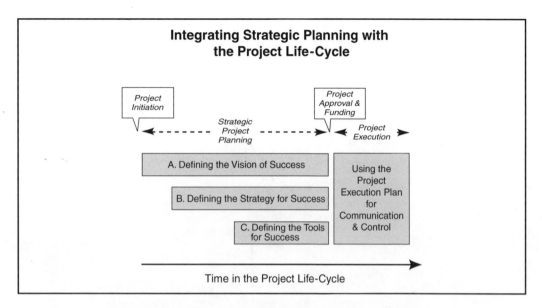

*Figure 5: Using The Strategic Project Planner
to develop a Project Execution Plan over the project's life cycle*

Strategic Project Planning is an ongoing process as the project progresses through the early stages. At first, the Project Execution Plan will be only partially complete, focusing on the project's business goals and project objectives (i.e., the Vision of Success). As the project becomes better defined, the risk management, organization, and contracting plans are developed (i.e., the Strategy for Success). Finally, a complete Project Execution Plan is presented as part of the request for final and full project funding. Once the project is funded and execution of the plan begins, the Project Execution Plan becomes the primary tool for communication, alignment, management, and control.

The Strategic Project Planner is a tool to help you develop and apply the Project Execution Plan that will help ensure project success.

III. How *The Strategic Project Planner* Enables the Implementation of the Project Management Institute's *Guide to the Project Management Body of Knowledge*

Project management is a relatively young and rapidly growing management discipline. The Project Management Institute (PMI®) is a leading professional association for the development and advancement of this important profession. (For more information visit *www.pmi.org*).

The Project Management Institute Standards Committee publishes the *Guide to the Project Management Body of Knowledge* (*PMBOK® Guide*), a book that has become a standard reference for many in the project management profession. The *PMBOK® Guide* is a key reference document for those who are practicing or learning project management, or qualifying for PMI's Project Management Professional (PMP) certification.

The *PMBOK® Guide* (see Section 4.1.3) describes the "Project Plan" (i.e., the Project Execution Plan) as follows:

> *The project plan is a formal, approved document used to manage and control project execution. . . . It commonly includes all of the following:*
>
> - *Project charter*
> - *Project Management strategy*
> - *Scope statement, which includes the project deliverables and the project objectives*
> - *Work Breakdown Structure*
> - *Cost estimates, scheduled start dates, and responsibility assignments*
> - *Performance measurement baselines for schedule and cost*
> - *Major milestones and target dates for each*
> - *Key or required staff*

- *Key risks, including constraints and assumptions, and planned responses for each*
- *Subsidiary management plans*
- *Open issues and pending decisions*

The *PMBOK® Guide* (see Section 4.1.2) describes a project planning methodology as follows:

A project planning methodology is any structured approach used to guide the project team during the development of a plan.

Your Strategic Project Planner is therefore a project planning methodology, used to create a Project (Execution) Plan. It is designed specifically to help you plan the implementation of the project management principles and Best Practices described in the *PMBOK® Guide*.

- *The Strategic Project Planner* will help your project team develop a Project Execution Plan that defines for each project *who* must do *what*, *why* it is important, and *when* it is required.

- The *PMBOK® Guide* will help your project team know *how* to carry out the critical project planning and execution functions required to implement the Project Execution Plan.

The complementary relationship between these two tools is illustrated by Figure 6.

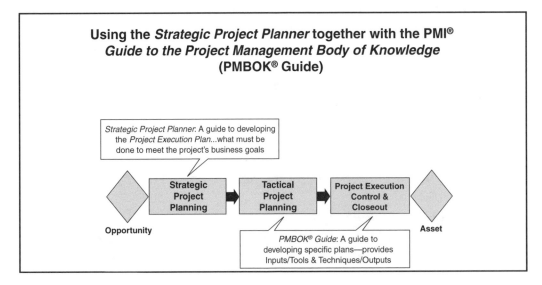

Figure 6: The Strategic Project Planner provides the framework for implementing the principles in the PMBOK® Guide

The *PMBOK® Guide* divides the subject of project management into nine Knowledge Areas:

- Project Integration Management
- Project Scope Management
- Project Time Management
- Project Cost Management
- Project Quality Management
- Project Human Resource Management
- Project Communications Management
- Project Risk Management
- Project Procurement Management

The Strategic Project Planner provides a framework for planning the project team's work in each of these areas. In some cases, a section in the Strategic Project Planner corresponds directly to a Knowledge Area in the *PMBOK® Guide*; in others, two or more sections of the Strategic Project Planner address a single Knowledge Area. The mapping of the *PMBOK® Guide* knowledge areas to sections in Westney's Strategic Project Planner is shown in Figure 7.

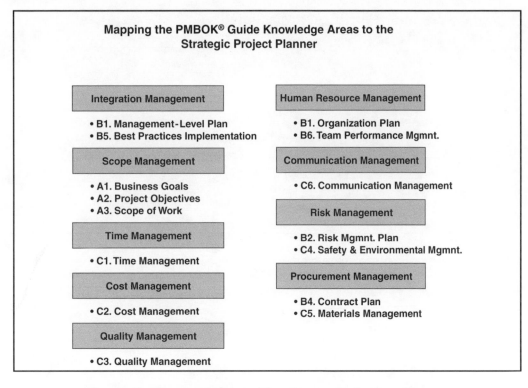

Figure 7: The modules in The Strategic Project Planner (indicated by bullets) support the planning of work in each Knowledge Area of the PMBOK® Guide (shown in boxes).

11

Note: The terms used in this book are consistent with the definitions pro-
       vided in the glossary of the PMBOK® Guide.

With *The Strategic Project Planner* in one hand and PMI's PMBOK® Guide
in the other, you have the tools you need to plan and execute successful
projects.

IV. How to Use *The Strategic Project Planner*

A. The Strategic Project Planning Modules

The Strategic Project Planner is divided into 15 planning modules, each of which corresponds to a step in the Strategic Project Planning Process as shown in Figure 4. These modules are as follows:

Part A: Defining the Vision of Success
> A1.   Business Goals
> A2.   Project Objectives and & Drivers
> A3.   Scope of Work

Part B: Defining the Strategy for Success
> B1.   Management-Level Plan
> B2.   Risk Management Plan
> B3.   Organization Plan
> B4.   Contract Plan
> B5.   Best Practices Implementation Plan
> B6.   Team Performance Management Plan

Part C: Defining the Tools for Success
> C1.   Time Management Plan
> C2.   Cost Management Plan
> C3.   Quality Management Plan
> C4.   Safety and Environmental Management Plan
> C5.   Materials Management Plan
> C6.   Communication Management Plan

Each module has its own section in the Strategic Project Planner.

IV. How to Use The
Strategic Project Planner

B. Open-Ended Questions Facilitate the Strategic Project Planning Process

Within each Strategic Project Planning module you will find an introduction page followed by a series of open-ended questions. These are the fundamental questions for which a Project Execution Plan must provide answers.

A sample page is shown below.

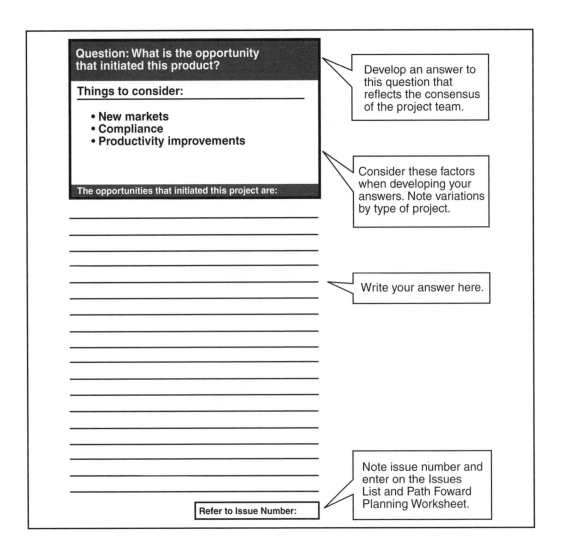

C. The "Things to Consider" Sections Are Industry-Specific Where Appropriate

The purpose of the "Things to Consider" section is to help to stimulate the strategic planning discussion for your project. Some considerations are important for any project, while others apply to specific industry groups or types of projects. The symbols indicate how to use the Things to Consider information.

Things to consider:

- **Information pertaining to all types of projects**

- **Information technology and telecommunications projects**

- **Energy projects—e.g., oil and gas, power generation**

- **Engineering and construction**

- **Transportation and infrastructure**

- **Manufacturing—e.g., refineries, chemical plant, pulp and paper, food, consumer products**

- **Pharmaceuticals and health care**

- **International projects**

```
Refer to
Issue #
```

D. Use the Issues List and Path Forward Planning Worksheet

As the Strategic Project Planning Process progresses, it is inevitable—indeed, desirable—that a number of significant issues are identified. It is very important to capture these issues (or action items) so they can be included in the Path Forward Plan.

Use the Issues List and Path Forward Planning Worksheet (Part VII) to capture these issues and action items as you work your way through the Strategic Project Planning Process.

V. Planning and Conducting the Strategic Project Planning Workshop

A project is a team effort, and a Project Execution Plan describes how that team effort will take place. It follows that the Strategic Project Planning Process is also a team effort, so that everyone on the team has input and is committed to the Project Execution Plan.

Successful project teams hold Strategic Project Planning workshops at critical points in the project. This section describes how to plan and facilitate a Strategic Project Planning workshop for your project.

A. Objectives of the Strategic Project Planning Workshop

The objectives of the workshop are usually to:

- Align project plans and objectives with business goals
- Develop the basis of a high-performing project team
- Identify planning issues and define a plan to resolve them
- Develop the basis for the Project Execution Plan
- Plan the application of Project Management Best Practices
- Get buy-in and commitment to the Project Execution Plan

B. Why Hold a Workshop?

Why hold a workshop? Why not simply nominate someone to write the Project Execution Plan? *Experience shows that the Strategic Project Planning Process itself is even more important than the Project Execution Plan that results.* Here is why:

- The workshop opens each team member's eyes to the perspective of the other participants

Project team members who have not participated in a Strategic Project Planning workshop often think that a planning workshop is unnecessary. They feel that they have a good understanding of the project's goals and objectives, scope of work, allocation of responsibilities, etc. The problem with this is that their understanding is apt to be quite different from that of other stakeholders, and even from that of the various customers whom the project must satisfy. It is easy to assume that everyone has all the information he or she needs, and is fully aligned, but that is hardly ever the case.

- The workshop harnesses the team's creativity

  Another reason for the workshop is the creative and problem-solving power of a "critical mass" of project participants. Not only can the group identify issues and problem areas that individual team members might not see, but it is also able to develop constructive ideas and plans to address them.

- The workshop achieves buy-in through participation

  Finally, it is essential that everyone participating in the project—either as a member of the project team or as a customer, supplier, advisor, or contractor—understand and be committed to the Project Execution Plan. What better way to achieve this than to have them participate in the development of that plan?

C. Who Should Attend the Workshop?

Participants in the Strategic Project Planning workshop should include:

- Members of the core project team, e.g.:
  - ➢ Project manager
  - ➢ Lead designer
  - ➢ Those in technical support disciplines
  - ➢ Those responsible for cost control
  - ➢ Planner/scheduler

- Members of the extended project team, e.g., in the following areas:
  - ➢ Legal
  - ➢ Purchasing
  - ➢ Health, safety, and environmental
  - ➢ Human Resources

- Customers, e.g.:
  - ➢ Marketing
  - ➢ Operations and maintenance
  - ➢ Technical services
  - ➢ Product support

- Management and administration

- Partners/investors, e.g.:
  - ➢ Other companies investing in the project
  - ➢ Lenders
  - ➢ Joint venture partners

- Key contractors and suppliers

Depending on the size of the project, this may result in the workshop having anywhere from 10 to 40 participants.

D. Who Facilitates the Workshop?

Most project teams prefer that someone from outside the project facilitate the workshop. This outside facilitator can ask the "dumb" questions that team members may be reluctant to ask. He or she can push the team to recognize flaws in their plan, or bring issues to the surface in a way that someone close to the project could not. And, of course, they can provide the service of planning, executing, and documenting the results of the workshop, when team members are loaded down with project tasks.

E. What Is the Agenda?

It is important to use the time of the large group wisely, and this usually means focusing on defining the Vision of Success (Part A) and defining the Strategy for Success (Part B). The facilitator should meet with the project manager and key participants prior to the workshop to identify priorities and design the agenda accordingly.

A typical agenda looks like the following:

- Introduction (Facilitator)
  Objectives of the workshop . . . introduction of the participants . . . explanation of the Strategic Project Planning Process.

V. Planning & Conducting the Strategic Project Planning Workshop

- Project Overview (Project Manager)
  Review of project history and current status. The purpose is to make sure that all participants have the information they need to be able to contribute to the meeting.

- Strategic Project Planning (Facilitator)
  Participants now address the open-ended questions in each section of the Strategic Project Planner, develop answers, and identify issues. This can be done in one large group or in breakout groups, with each breakout assigned a module. The breakout teams develop answers to the assigned questions, define issues, and then present their results to the entire group.

  For most projects, the modules that are most appropriate to work on in the workshop are:

  - Part A: Defining the Vision of Success
    A1. Business Goals
    A2. Project Objectives and Drivers
    A3. Scope of Work

  - Part B: Defining the Strategy for Success
    B1. Management-Level Plan
    B2. Risk Management Plan
    B3. Organization Plan
    B4. Contract Plan
    B5. Best Practices Implementation Plan
    B6. Team Performance Management Plan

  - Part C, Defining the Tools for Success, can usually be done effectively with smaller, specialized groups in separate meetings after the workshop.

- Review and Consolidation of Issues and Action Items (Facilitator)

- Path Forward Planning (Facilitator)

F. How Does the Planning Get Done?

Most project teams prefer that each participant in the workshop have his or her own copy of the Strategic Project Planner. They use the Planner during the workshop, and later as a personal planning tool as the project progresses.

As each section of the Strategic Project Planner is addressed, participants take notes on the work of their own breakout team as well as on the results presented by the other breakout groups.

Planning results can be captured on flipcharts, on transparencies, or on a personal computer. It is particularly effective to project the computer screen so that everyone can see the answers being developed for each of the planning questions. The projector can also be used to develop the Issues List and Path Forward Action Plan. Not only does this ensure that the workshop information is captured, it enhances the workshop by getting everyone to focus on the questions and issues at hand. The result is an electronic Project Execution Plan that can easily be e-mailed to all participants—both the original plan and subsequent updates.

V. Planning & Conducting
the Strategic Project
Planning Workshop

VI. *The Strategic Project Planner*

Part A: Defining the Vision of Success

No project can succeed unless the vision of success is clear to everyone. Yet too often we take for granted that everyone knows the goals of the project and is committed to them. In fact, it is rare that all project participants have the same goals. For example:

- The project manager, knowing the punishment that awaits those whose projects overrun, wants to be sure the final cost is within the budget.

- Operations wants to be sure that the project delivers to them a system or facility that is reliable and easy to operate and maintain.

- Business Unit Management wants the project to be done with a minimum of risks and resources, even though lack of resources is the surest way to increase the risk of failure.

- Corporate Management wants to minimize the amount of effort and money spent before the project is approved, even though lack of scope definition is the surest way to ensure that estimates and schedules will be unrealistic.

- Contractors and suppliers want to ensure their own profits, a goal that does not necessarily align with minimizing project costs.

Defining the Vision of Success is therefore a process of:

- Clarifying the business goals that define overall project success

- Translating these business goals into project objectives and drivers

- Defining the scope of work that will meet the business goals and satisfy the project objectives.

This process is illustrated in the following diagram below.

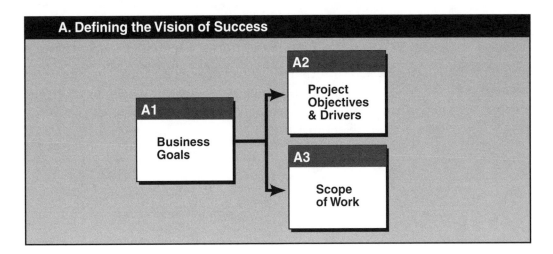

A1: Business Goals

Objective: Identify the business goals that define project success.

A successful project is one that meets or exceeds its business goals. Therefore, the first and most important step in Strategic Project Planning is to make sure the project team understands what those business goals are.

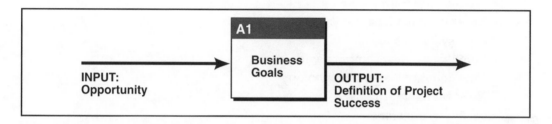

Business goals are usually set by someone outside the project team, although at times the project manager may be involved. Therefore, the objective of this step in the Strategic Project Planning Process is to make sure that everyone on the project team understands and is committed to achieving these goals. It is important that those responsible for business goals be represented in the discussion, to ensure that the project team has "heard them correctly."

Most teams find that, even when the business goals (e.g., "make a profit") seem obvious, it is not long before the true complexity and issues emerge. There are usually a number of business goals for the project, and they may be somewhat contradictory, so a thorough discussion of the questions in this module is recommended. Teams should also remember that business goals may change over time, and such changes may not have been communicated to the project team. Goals should therefore be updated periodically.

PMBOK® Guide reference:
5.1: Initiation

A1: Business Goals

a. What is the opportunity that initiated this project?

Things to consider:

- New markets
- Compliance with regulations
- Productivity improvements
- New systems technology
- New systems capabilities
- Compliance with company-wide systems initiatives
- New communications technology
- Need for new facilities
- Facility expansions
- Turnarounds/outages
- Government funding
- Repairs or upgrades to existing facilities

- Scheduled maintenance of existing facilities
- Need to improve/increase utilities or infrastructure
- Need to expand capacity
- New drug discovery
- Compliance with Good Manufacturing Practice

- Developing markets
- New trade treaties
- Economic conditions

The opportunity that initiated this project is:

_____

_____

_____

_____

_____

_____

_____

_____

| Refer to |
| Issue # |

A1: Business Goals

b. What is the asset this project will create?

Things to consider:

- Capital asset
- New product
- New knowledge
- New information management capability
- New communication capability
- New work processes
- New oil or gas reservoir
- New production or transportation facility
- New power generation facility
- New capital construction
- Completed design package

- Permanent infrastructure
- Expansion to existing facilities

- New manufacturing facility
- New equipment or utilities
- Completed turnaround
- New product
- New research facility
- Testing or research results
- Presence in an overseas location

The asset this project will create is:

_____

_____

_____

_____

_____

_____

_____

Refer to Issue #

A1: Business Goals

c. What is the economic justification for this project?

Things to consider:

- Return on investment
- Increased market share
- Compliance with regulations
- Increased efficiency
- Introduction of new products or technology
- Development of new markets
- Diversification

The economic justification for this project is:

_____

_____

_____

_____

_____

_____

_____

_____

_____

_____

_____

_____

_____

_____

_____

_____

_____

_____

_____

**Refer to
Issue #**

A1: Business Goals

d. What are the economic sensitivities for this project?

Things to consider:

- What is an acceptable cost increase to improve the schedule?
- Market conditions
- Capital cost
- Life-cycle cost
- International business conditions
- Currency fluctuations

The economic sensitivities for this project are:

_____

_____

_____

_____

_____

_____

_____

_____

_____

_____

_____

_____

_____

_____

_____

_____

_____

_____

_____

_____

Refer to Issue #

A1: Business Goals

e. What overall corporate strategic goals will impact this project?

Things to consider:

- New sources of revenue
- Product or market diversification
- Environmental policy and goals
- Safety policy and goals
- Cash-flow and budget constraints
- International expansion
- Diversification

The corporate strategic goals that impact this project are:

_____

_____

_____

_____

_____

_____

_____

_____

_____

_____

_____

_____

_____

_____

_____

_____

_____

_____

_____

_____

(Note: corporate strategic
goals apply to *all* projects.)

| Refer to
Issue # |

A1: Business Goals

f. How do marketing goals impact this project?

Things to consider:

- Date of first production or system readiness
- Quality specification
- Reliability
- Contractual commitments

Marketing goals have the following impact on this project:

_____

_____

_____

_____

_____

_____

_____

_____

_____

_____

_____

_____

_____

_____

_____

_____

_____

_____

_____

_____

_____

_____

_____

| Refer to |
| Issue # |

A1: Business Goals

g. What are this project's goals with respect to the demonstration or application of new technology?

Things to consider:

- Company image
- Opportunity to exploit internal research & development
- Meeting special customer requirements
- Gaining competitive advantage
- New technology may be essential to achieving the business goals
- Need to demonstrate the feasibility of proprietary new technology
- Communication technology and systems
- Internet technology
- Hardware technology
- Software technology
- Subsea technology
- Drilling technology
- 3D seismic technology
- Hydrocarbon processing technology
- Hydrocarbon process controls technology
- Computerized design methods
- New construction methods
- New materials technology
- Traffic control technology
- Vehicle technology

- Process technology
- Equipment technology
- Materials technology
- Production controls technology
- Manufacturing methods
- Pharmaceutical science

- Technologies that may be new to host country
- Technologies from host country that may be new to owner

This project's goals with respect to new technology are:

_____

_____

_____

_____

| Refer to |
| Issue # |

A1: Business Goals

h. What will be the nature and extent of pre-investment for additional future capabilities?

Things to consider:

- **Future capacity increases**

- **Ability to accommodate future technical developments**

- **Secondary or tertiary recovery**

- **Growth in traffic capacity**

- **Future variations in product specifications**
- **Future variations in feedstock specifications**
- **Flexibility to manufacture alternative products**
- **Infrastructure to meet future needs**
- **Training facilities**

This project includes or may include pre-investment for:

_____

_____

_____

_____

_____

_____

_____

_____

_____

_____

_____

**Refer to
Issue #**

A1: Business Goals

i. What other projects may impact, or be impacted by, this one?

Things to consider:

- Projects that depend on this one to be complete
- Projects providing facilities or services that this project requires

The following projects impact or are impacted by this project:

_____

_____

_____

_____

_____

_____

_____

_____

_____

_____

_____

_____

_____

_____

_____

_____

_____

_____

_____

_____

_____

_____

_____

_____

Refer to **Issue #**

A1: Business Goals

j. What is the priority of this project relative to other similar projects in the same timeframe?

Things to consider:

- Priority based on return on investment
- Priority based on contractual commitments
- Priority based on statutory requirements
- Priority based on customer needs

This project's priority relative to other projects in the same timeframe is as follows:

_____

_____

_____

_____

_____

_____

_____

_____

_____

_____

_____

_____

_____

_____

_____

_____

_____

_____

_____

_____

**Refer to
Issue #**

A1: Business Goals

k. How will tax-related financial strategies impact the project objectives?

Things to consider:

- Date facilities placed in service
- Tax relief on purchased items
- Tax credits for capital or operating costs
- Host country tax regulations
- Impact of shared ownership with nationally owned company

Tax-related financial strategies can or will impact the project as follows:

_____

_____

_____

_____

_____

_____

_____

_____

_____

_____

_____

_____

_____

_____

_____

_____

_____

_____

_____

_____

| Refer to |
| Issue # |

A1: Business Goals

I. What are the issues with respect to the project's business goals?

Enter each issue on the Issues List and Path Forward Planning Worksheet.

A2: Project Objectives and Drivers

Objective: Develop project objectives that will fulfill the business goals

Once the business goals are understood by the project team, the next step is to translate these goals into project objectives. Project objectives set the standard for performance against the variables that a project team can control. For example, profitability is usually a business goal, and a key element in product profitability is the product price—something that the project team has limited or no control over. However, the cost of the manufacturing facility is also a key profitability variable, and one that becomes a project objective since the project team has control over it.

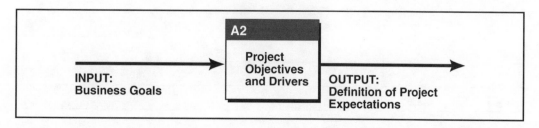

It is helpful to remember that project objectives are not really fixed. No project manager can plan a project to finish exactly (to the dollar) on budget or exactly (to the hour) on schedule. Much as we would like to think otherwise, project objectives have some tolerance around them, and the team needs to understand how to make tradeoffs between one objective (such as cost) and another (such as time).

PMBOK® Guide reference:
5.1 Initiation
5.2 Scope Planning

A2: Project Objectives and Drivers

a. What is the current project cost estimate and what is its accuracy?

Things to consider:

- Estimate accuracy (i.e., range of probable costs)
- Reconciliation with previous estimates
- Basis of the estimate—scope, plan, cost data and estimating methods
- Amount of estimate contingency
- Probability that the actual final cost will exceed the target estimate

The value and accuracy of the current cost estimate are:

_____

_____

_____

_____

_____

_____

_____

_____

_____

_____

_____

_____

_____

_____

_____

_____

_____

_____

_____

_____

_____

| Refer to |
| Issue # |

A2: Project Objectives and Drivers

b. What is the maximum project cost that will allow the project's business goals to be met?

Things to consider:

- Sensitivity of return on investment to capital cost
- Project cost associated with minimum acceptable return on investment

The maximum project cost that will still allow the business goals to be met is:

_____

_____

_____

_____

_____

_____

_____

_____

_____

_____

_____

_____

_____

_____

_____

_____

_____

_____

_____

_____

_____

_____

Refer to Issue #

A2: Project Objectives and Drivers

c. What is the current target completion date and what is its accuracy?

Things to consider:

- Optimal completion date for maximum return-on-investment
- Optimal completion date to meet contractual requirements
- Scheduling accuracy (i.e., range of possible end dates)
- Reconciliation with previous schedules
- Amount of schedule contingency
- Probability that the actual final end date will be later than the target completion date

The value and accuracy of the current target completion date are:

_____
_____
_____
_____
_____
_____
_____
_____
_____
_____
_____
_____
_____
_____
_____
_____
_____
_____
_____
_____
_____

| Refer to |
| Issue # |

A2: Project Objectives and Drivers

d. What is the latest completion date that will allow the project's business goals to be met?

Things to consider:

- Completion date to satisfy minimum return on investment
- Sensitivity of return on investment to completion date
- Penalties for failing to meet schedule commitments

The latest completion date that will allow the project's business goals to be met is:

_____

_____

_____

_____

_____

_____

_____

_____

_____

_____

_____

_____

_____

_____

_____

_____

_____

_____

_____

_____

_____

Refer to Issue #

A2: Project Objectives and Drivers

e. What quality requirements are needed to meet the project's business goals?

Things to consider:

- Conformance to customer requirements
- Conformance to standards and codes
- Minimum rework
- Minimum rejection rate
- Performance or operational standards
- System performance specifications
- Hardware specifications
- User perceptions of quality and functionality
- Amount of rework due to design and/or construction errors

- Conformance of product to specifications

- Conformance of manufacturing procedures to GMP or other requirements

- Prevailing quality standards at the project location

The quality requirements are as follows:

_____

_____

_____

_____

_____

_____

_____

_____

_____

_____

| Refer to |
| Issue # |

A2: Project Objectives and Drivers

f. What are the project's safety objectives?

Things to consider:

- Safety during the project
- Safety after the project is complete (i.e., operational safety)
- Statutory requirements
- Company goals and standards
- Safety of the equipment and facilities
- Safety of data

- Safety design for emergency shutdown
- Incident prevention
- Safety of the facilities
- Design for safe construction
- Construction safety performance
- Use of safety plans and measurements
- OSHA and other regulatory requirements
- Site security
- Operational safety
- Regulatory safety requirements
- Protection from sabotage, etc.
- Operational safety

- Product safety

- Local safety standards and practices
- Prevailing unsafe conditions and practices
- Need for safety training

The project's safety objectives are:

_____

_____

_____

_____

| Refer to |
| Issue # |

A2: Project Objectives and Drivers

g. What are the project's environmental objectives?

Things to consider:

- Federal, state, and local statutory requirements
- Air emissions
- Water contamination
- Impact on wildlife
- Visual impact
- Waste disposal
- Avoidance of, or reaction to, oil spills

- Permitting
- Soil remediation

- Host-country environmental standards and requirements

The project's environmental objectives are:

_____

_____

_____

_____

_____

_____

_____

_____

_____

_____

_____

_____

_____

_____

| Refer to |
| Issue # |

A2: Project Objectives and Drivers

h. What are the project's objectives with respect to life-cycle costs?

Things to consider:

- Capital cost
- Operating costs
- Maintenance costs
- Utility costs
- Taxes
- Dismantling costs vs. salvage value

The project's life-cycle cost objectives are:

_____

_____

_____

_____

_____

_____

_____

_____

_____

_____

_____

_____

_____

_____

_____

_____

_____

_____

_____

_____

_____

_____

| Refer to |
| Issue # |

A2: Project Objectives and Drivers

i. What public-relations or community-relations objectives must be met?

Things to consider:

- Company image
- Local image
- Impact of project on local community
- Domestic content
- Impact on local infrastructure
- Need to develop local contractors and suppliers

The project's public- and community-relations objectives are:

_____

_____

_____

_____

_____

_____

_____

_____

_____

_____

_____

_____

_____

_____

_____

_____

_____

_____

_____

**Refer to
Issue #**

A2: Project Objectives and Drivers

j. How are the project's drivers to be prioritized in order to meet the business goals?

Things to consider:

- Typical project drivers include:
  - ➢ Capital cost
  - ➢ Schedule
  - ➢ Quality
  - ➢ Safety
  - ➢ Life-cycle cost
  - ➢ Reliability

The ranking of the project's drivers is as follows:

_____

_____

_____

_____

_____

_____

_____

_____

_____

_____

_____

_____

_____

_____

_____

_____

_____

_____

_____

_____

Note: Drivers provide the basis for daily planning and decision-making; they ensure that all project actions support the project's business goals.

Refer to Issue #

A2: Project Objectives and Drivers

k. What cash-flow constraints may impact the project?

Things to consider:

- Budget constraints
- Phased funding
- Project financing
- Funding by partners
- Time required to gain funding approvals

The following cash-flow constraints could impact the project:

_____

_____

_____

_____

_____

_____

_____

_____

_____

_____

_____

_____

_____

_____

_____

_____

_____

_____

_____

_____

_____

| Refer to |
| Issue # |

A2: Project Objectives and Drivers

I. What operational objectives drive the project's design, plans, and schedule?

Things to consider:

- Coordinating startup with existing operations
- Coordinating startup with other new facilities and systems
- Timing for startup

The project's design, plans, and schedule must reflect the following operational objectives:

**Refer to
Issue #**

A2: Project Objectives and Drivers

m. What are the issues with respect to the project's Objectives & Drivers?

Enter each issue on the Issues List and Path Forward Planning Worksheet.

A3: Scope of Work

Objective: Define the scope of work that will fulfill the business goals

The failure to define and control a project's scope of work is unquestionably the most frequent cause of failure of projects to meet their objectives. But scope of work is not easy to define. Often, those who have the ability to change the scope (i.e., "customers") are not available during the early planning. And full scope definition requires that a project be well-advanced for sufficient definition to be established.

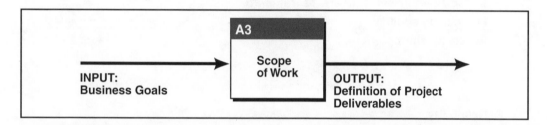

A thorough discussion of the questions in this section will help ensure that all scope issues are defined. This allows the Path Forward Plan developed at the end of the strategic planning session to focus on the most critical scope issues.

PMBOK® Guide reference:
4.3 Overall Change Control
5.2 Scope Planning
5.3 Scope Definition
5.4 Scope Verification
5.5 Scope Change Control

53

A3: Scope of Work

a. What is the functional performance required to meet the business goals?

Things to consider:

- Customer requirements
- Technical constraints

- User functionality
- Capacity—number of users, storage, etc.
- Processing speed
- Reservoir characteristics

- Load capacity
- Reliability

- Production capacity
- Feedstock specification
- Product specification
- Operating and maintenance philosophy
- GMP compliance

The *functional scope* of the project is to provide the following performance or functionality:

(Note: the "functional scope" defines what the asset will do.)

Refer to Issue #

A3: Scope of Work

b. What is the current technical definition of the scope of work?

Things to consider:

- Design standards
- Technical definition
- Operating philosophy
- Equipment specification
- Systems design
- Communications technology

- Instruments and controls
- Sparing philosophy

- Layout and location

- Process design
- Automation
- Plot plan
- Manufacturing process
- Packaging
- Materials specification
- Local design and operations, and maintenance practices

The current technical definition of the scope of work is as follows:

_____

_____

_____

_____

_____

_____

_____

_____

(Note: the "technical scope"
describes the design of the asset.)

| Refer to |
| Issue # |

A3: Scope of Work

c. To what extent and in what way is new technology included in the technical scope?

Things to consider:

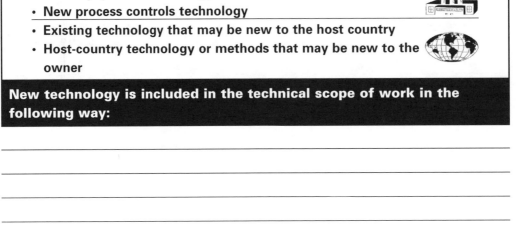

- Design uncertainties caused by new technology
- Potential impact of new technology on the design and scope
- Potential impact of new technology on risks
- Potential impact of new technology on cost and schedule
- Potential impact of new technology on specialist resource requirements
- New communication technology
- New hardware technology
- New software technology
- New seismic technology
- New drilling technology
- Deepwater drilling and production
- New construction technology
- New materials technology
- New design methods and technology
- New vehicle technology
- New navigation technology
- New communication technology
- New process technology
- New equipment technology
- New process controls technology
- Existing technology that may be new to the host country
- Host-country technology or methods that may be new to the owner

New technology is included in the technical scope of work in the following way:

| Refer to |
| Issue # |

A3: Scope of Work

d. What are the physical deliverables that the project must create?

Things to consider:

- Use a Work Breakdown Structure to define the project deliverables (see Question "I")
- Describe deliverables as specifically as possible
- Specify what deliverables are *not* included in the scope
- Operational systems
- Manufacturing facilities

- Production facilities
- Pipelines
- Storage facilities
- Completed construction

- Bridges, roads, tunnels
- Airport facilities
- Marine facilities
- Chemical manufacturing facilities
- Refining facilities
- Product manufacturing facilities
- Research facilities
- Product packaging facilities
- Clinical testing results
- Project facilities
- Infrastructure

The project must create the following deliverables:

_____

_____

_____

_____

_____

_____

(Note: The physical deliverables usually describe the asset that the project creates.)

| Refer to |
| Issue # |

A3: Scope of Work

e. What is the design philosophy for this project?

Things to consider:

- Impact of design philosophy on business goals
- Useful economic life
- Design life
- Reliability/redundancy/sparing philosophy
- Technical obsolescence
- Operational flexibility
- Reliability
- Reservoir life

- Contract design practices and standards

- Production flexibility
- Corporate design standards
- Local standards
- Product life
- Mandated design standards

- International standards
- Local standards
- Need to meet local operating requirements

The project's design philosophy is as follows:

Refer to
Issue #

A3: Scope of Work

f. What are the overall activities that are included in the project's scope of work?

Things to consider:

- **Activities described in the Management-Level Plan (see Section B1)**
- **Owner and contractor activities**

The project's scope of work includes the following major activities:

_____

_____

_____

_____

_____

_____

_____

_____

_____

_____

_____

_____

_____

_____

_____

_____

_____

_____

_____

_____

_____

_____

Refer to Issue #

A3: Scope of Work

g. What necessary or desirable scope is currently excluded from the project's scope?

Things to consider:

- Necessary or desirable activities to be performed by others
- Necessary or desirable deliverables to be provided by others
- Necessary or desirable facilities to be provided by others
- Necessary or desirable functionality to be provided by others

The project's functional, technical, physical, and activity scope of work currently excludes:

_____

_____

_____

_____

_____

_____

_____

_____

_____

_____

_____

_____

_____

_____

_____

_____

_____

_____

_____

_____

_____

Refer to **Issue #**

A3: Scope of Work

h. What is the likelihood and extent of potential future changes to project scope?

Things to consider:

- Technical uncertainty
- Degree of customer commitment to the current scope definition
- Effectiveness of the scope management program
- Uncertainty due to design complexity
- Uncertainty due to the "retrofit" factor if the project is an addition or modification to an existing system or facility
- Uncertainty regarding interfaces
- Organizational tension between the project team and the customer
- Lack of effort to date to fully define scope
- Changes in prevailing or expected technology

- Impact of other projects in same area
- Changes to production agreements
- Changes in reservoir characteristics
- Increases in required capacity
- Added scope to repair or upgrade existing facilities

- New research developments
- Results of clinical pharmacology

- New requirements for local infrastructure
- New partners

Potential changes to the scope of work are:

_____

_____

_____

_____

_____

| Refer to |
| Issue # |

A3: Scope of Work

i. How well does the current cost estimate reflect the known or anticipated project scope of work?

Things to consider:

- Timing of the current cost estimate (when was it done?)
- Scope definition on which the estimate was based (how does it compare with the current scope?)
- Estimate assumptions (are they still valid?)
- Degree to which the design and scope have progressed since the estimate was done
- Plans for estimate updates
- Estimating methods used to calculate the cost of scope that is currently undefined
- Amount of scope that is still undefined

Differences between the scope on which the current estimate is based and the current scope of work are as follows:

_____

_____

_____

_____

_____

_____

_____

_____

_____

_____

_____

_____

_____

_____

_____

_____

_____

Refer to Issue #

A3: Scope of Work

j. How well will contractor's scope be defined when contracts are let?

Things to consider:

- Type of contract (e.g., lump sum requires comprehensive scope definition; reimbursable contracts do not)
- Timing of contract award (how well will scope be defined by then?)
- Responsibility for defining contract scope

The degree of definition of contractor's scope at the time of contract award will be as follows:

_____

_____

_____

_____

_____

_____

_____

_____

_____

_____

_____

_____

_____

_____

_____

_____

_____

_____

_____

_____

_____

_____

| Refer to |
| Issue # |

A3: Scope of Work

k. How will future changes to scope be controlled?

Things to consider:

- What is the change management procedure?
- Who has responsibility for the cost and schedule impact of scope changes?
- How will the cost and schedule impact of scope changes be estimated?
- Who approves scope changes?
- What criteria will be used to justify and approve scope changes?

Potential future changes to the scope of work will be controlled as follows:

_____
_____
_____
_____
_____
_____
_____
_____
_____
_____
_____
_____
_____
_____
_____
_____
_____
_____
_____
_____

**Refer to
Issue #**

A3: Scope of Work

I. What is the Work Breakdown Structure for the Project?

Sample Work Breakdown Structure
Project: A New Residential Home

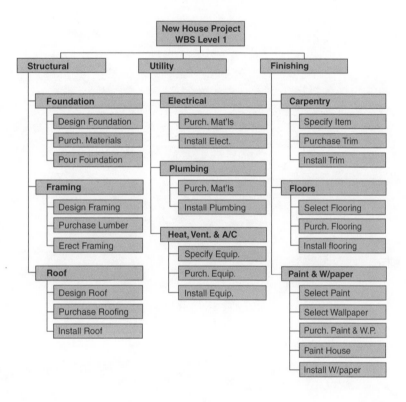

Note: The Work Breakdown Structure (WBS) is a hierarchical representation of the project's scope of work. It is very useful to define functional, physical, and activity scope. Once the WBS is complete, it can be used to guide the development of the Organization Plan, the Contract Plan, and the Activity Plan. Refer to the PMBOK® Guide, Section 5, for more information on Work Breakdown Structure.

A3: Scope of Work

m. What are the issues with respect to the project's scope of work?

Enter each issue on the Issues List and Path Forward Planning Worksheet.

VI. *The Strategic Project Planner*

Part B: Defining the Strategy for Success

The Vision of Success, described in Part A, focuses on the *business goals* and *project objectives*. Once that has been established, you are ready to proceed to the Strategy for Success, which focuses on *risks* and *resources*.

A strategy begins with a Management-Level Plan that identifies the project's key decision points, milestones, and deliverables at each phase in the plan. This serves as a basis for identifying risks and developing a Risk Mitigation Plan. Project resources are addressed at a strategic level by the Organization Plan, the Contract Plan, and the Team Performance Management Plan.

The planning process to define the Strategy for Success is shown in the following diagram.

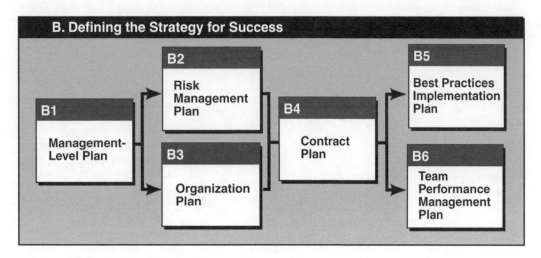

B1: Management-Level Plan

Objective: Develop the Management-Level Plan to meet project objectives

Before a comprehensive project plan can be developed, it is very useful to have a Management-Level Plan. The Management-Level Plan specifies:

- Where the major decision points are
- What the phases of the project are
- What the major milestones are
- What the overall level of resource requirements will be

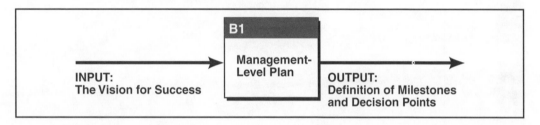

Many companies are applying the Project Management Best Practice of using a common "stage-gate" process for all projects. In this practice, projects are divided into "stages" (or "phases") which are separated by decision points, or "gates." This provides a ready-made format for the Management-Level Plan and ensures discipline as the project progresses.

PMBOK® Guide reference:
2.1 Project Phases and the Project Life Cycle
4.1 Project Plan Development
5.1 Initiation

Note: Worksheets for developing the management-level project plan and the key milestones are provided on the following pages.

Management-Level Project Planning Worksheet	Phase I	Phase II	Phase III	Phase IV	Phase V
What is the decision to be made at the end of this phase?					
Who will make the decision?					
What are the deliverables required for the decision?					
What are the critical tasks for this phase?					
What resources will be required?					

B1. Management-Level Plan

Management-Level Project Planning Worksheet

	Phase I	Phase II	Phase III	Phase IV	Phase V
What is the decision to be made at the end of this phase?					
Who will make the decision?					
What are the deliverables required for the decision?					
What are the critical tasks for this phase?					
What resources will be required?					

Management-Level Project Planning Worksheet	Phase I	Phase II	Phase III	Phase IV	Phase V
What is the decision to be made at the end of this phase?					
Who will make the decision?					
What are the deliverables required for the decision?					
What are the critical tasks for this phase?					
What resources will be required?					

Milestone Analysis Table

Schedule Milestone *(or Activity)*	Required Date *(or Duration)*	Likelihood It Can Be Met (High, Avg., Low)	Discussion of Criticality, Restraints, Issues, Risks, etc.

B1. Management-Level Plan

Milestone Analysis Table			
Schedule Milestone *(or Activity)*	Required Date *(or Duration)*	Likelihood It Can Be Met (High, Avg., Low)	Discussion of Criticality, Restraints, Issues, Risks, etc.

Milestone Analysis Table			
Schedule Milestone *(or Activity)*	Required Date *(or Duration)*	Likelihood It Can Be Met (High, Avg., Low)	Discussion of Criticality, Restraints, Issues, Risks, etc.

B1. Management-Level Plan

B1: Management-Level Plan

a. What are the project *phases* that describe the work between decision points?

Things to consider:

- Initiation phase—begins development of the opportunity into various strategies
- Feasibility phase—develops alternative strategies until the optimal one is chosen
- Definition phase—defines the preferred strategy to the point at which an estimate of time, cost, risk, and profitability can be made with sufficient accuracy to justify full project funding
- Execution phase—creates the asset
- Evaluation phase—the asset begins producing revenue and the team evaluates how well the project development and execution process worked

Enter the project phases on the Management-Level Project Planning Worksheet

_____

_____

_____

_____

_____

_____

_____

_____

_____

_____

_____

_____

_____

_____

_____

_____

| Refer to |
| Issue # |

B1: Management-Level Plan

b. What are the *key decision points* in the project life cycle?

Things to consider:

- Decision to fund feasibility studies
- Decision to fund project definition
- Decision to fund project execution
- Decision to accept project and begin evaluation
- Interim strategic decisions
- Decisions regarding technology selection

- Decisions to conclude production agreements
- Financing decisions
- Decisions to invest in seismic or drilling programs
- Decisions to award key contracts

- Process-selection decisions
- Site-selection decisions

- Decisions to commit to host-country authorities
- Financing decisions

Enter the key project decision points on the Management-Level Project Planning Worksheet

_____

_____

_____

_____

_____

_____

_____

_____

_____

_____

_____

| Refer to |
| Issue # |

B1: Management-Level Plan

c. Who are the *decision-makers* at each decision point?

Things to consider:

- Management level required to authorize funds
- Customer participation
- Partner participation
- Technical participation
- Operational participation
- Other stakeholders

Enter the key decision-makers on the Management-Level Project Planning Worksheet

_____

_____

_____

_____

_____

_____

_____

_____

_____

_____

_____

_____

_____

_____

_____

_____

_____

_____

_____

_____

_____

_____

Refer to Issue #

B1: Management-Level Plan

d. What are the *deliverables* required to support the decision at each decision point?

Things to consider:

- **Project economics and justification**
- **Cost estimate**
- **Project Execution Plan (i.e., Strategic Project Plan)**
- **Task plan and schedule**
- **Resource plan and schedule**
- **Scope definition**
- **Risk analysis**

Enter the decision-support deliverables at each decision point on the Management-Level Project Planning Worksheet

Note: the more complex, costly, risky, or difficult the decision, the more decision-support deliverables are likely to be required.

**Refer to
Issue #**

B1: Management-Level Plan

e. What *tasks* are required to create these decision-support deliverables?

Things to consider:

- Preparation of economic analysis
- Scope definition
- Interim decision-making
- Strategic Project Planning
- Design reviews
- Project risk analysis
- Task and resource planning and scheduling
- Cost estimating
- Sensitivity studies
- Scenario planning
- Decision analysis

Enter the tasks required to create the decision-support deliverables on the Management-Level Project Planning Worksheet

_____

_____

_____

_____

_____

_____

_____

_____

_____

_____

_____

_____

_____

_____

_____

_____

_____

_____

**Refer to
Issue #**

B1: Management-Level Plan

f. What *resources* are required to create these decision-support deliverables?

Things to consider:

- Engineering—for design and scope definition
- Operations and maintenance—to define requirements and support decision-making
- Marketing and customer—to define requirements and support decision-making
- Research & development
- Corporate planning—to define business goals and calculate economics
- Project management—to develop strategies, plans, schedules, and cost estimates
- Experts on conditions in the host country

Enter the resources required to complete the tasks to create the decision-support deliverables on the Management-Level Project Planning Worksheet

_____
_____
_____
_____
_____
_____
_____
_____
_____
_____
_____
_____
_____
_____
_____

| Refer to |
| Issue # |

B1: Management-Level Plan

g. What are the critical milestones within each phase?

Things to consider:

- Funding approvals
- Project completion date
- Key decision points
- Contractual requirements
- Date of equipment availability
- Cutover dates

- Date of start of drilling
- Compliance date
- Date for turnarounds
- Weather windows
- Seasonal construction conditions
- Permitting timing and dates
- Legislative commitments
- Commitments to community

- Date of first production
- Shutdown dates
- Sales commitments
- Key testing dates
- Validation schedule

- Shipping requirements
- Key political dates

Enter the critical milestones on the Milestone Analysis Table

Refer to
Issue #

B1: Management-Level Plan

h. What is the overall probability (e.g., high, medium, low) that each critical milestone will be met?

Things to consider:

- Optimism or pessimism in setting deadlines
- Past experience
- Forecast for scope or design changes
- Performance data from past projects
- Lead times required for funding approvals
- Schedule–related risks (see B2, Risk Management Plan)

Enter the probability of meeting each critical milestone on the Milestone Analysis Table

_____
_____
_____
_____
_____
_____
_____
_____
_____
_____
_____
_____
_____
_____
_____
_____
_____
_____
_____

Refer to
Issue #

B1. Management-Level Plan

B1: Management-Level Plan

i. What are the issues with respect to the Management-Level Plan?

Enter each issue on the Issues List and Path Forward Planning Worksheet

B2: Risk Management Plan

Objective: Identify risks to project success and develop plans to mitigate them

Project risk may well be the last truly neglected project control variable. By "project risk" we mean the probabilities that the project will fail to meet its objectives. The degree of risk and uncertainty affects the accuracy of our estimates of time and cost, and the likelihood of an overrun. Yet most projects still give only a token effort to analyzing these risks.

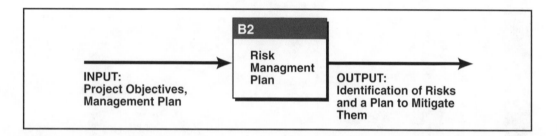

Risks can be mitigated, and this section also provides a simple technique for identifying risks, assessing their potential impact, and then developing a plan to mitigate them.

PMBOK® Guide reference:
11.1 Risk Identification
11.2 Risk Quantification
11.3 Risk Response Development
11.4 Risk Response Control

Note: Risk Mitigation Planning charts are provided on the following pages.

Risk Mitigation Planning Chart

Risk Identification			Risk Mitigation				
Describe Risk	Prob. of Occurrence (High, Med., Low)	Potential Impact (H,M,L)	Describe Action	Cost to Mitigate (H,M,L)	Prob. of Success (H,M,L)	Do It? (Yes/No)	Who Leads?	By When?

Risk Mitigation Planning Chart

Risk Identification			Risk Mitigation					
Describe Risk	Prob. of Occurrence (High, Med., Low)	Potential Impact (H,M,L)	Describe Action	Cost to Mitigate (H,M,L)	Prob. of Success (H,M,L)	Do It? (Yes/No)	Who Leads?	By When?

B2. Risk Management Plan

Risk Mitigation Planning Chart

	Risk Identification			Risk Mitigation					
Describe Risk	Prob. of Occurrence (High, Med., Low)	Potential Impact (H,M,L)	Describe Action	Cost to Mitigate (H,M,L)	Prob. of Success (H,M,L)	Do It? (Yes/No)	Who Leads?	By When?

B2: Risk Management Plan

a. What are the external risks to project success? (see checklist on next page)

Things to consider:

- Economic fluctuations
- Government actions, changes in regulations
- Financing difficulties
- Changes in market conditions for key project materials or services
- Extraordinary weather conditions
- Legal difficulties

- Changes in technology
- Changes in regulations
- Competing systems
- Extraordinary weather conditions during drilling, offshore installation, or hookup
- Unforeseeable reservoir conditions

- Extraordinary weather conditions
- Unforeseeable changes in labor conditions
- Unforeseeable site difficulties
- Environmental compliance issues
- Community resistance and interference
- Environmental impacts

- Unforeseeable changes in technology
- Interference from Operations or Maintenance

- Competing products in same timeframe
- Unforeseeable problems with government approvals

- Political risks in host country
- Internal unrest
- External conflicts
- Currency fluctuations
- Changes in tax or other regulations

Enter the external project risks on the Risk Mitigation Planning Chart

External Risks: The company cannot control the occurrence of these risks, but it can mitigate their impact.

Refer to Issue #

B2: Risk Management Plan

b. What are the internal risks to project success?

Things to consider:

- Poor scope definition and control
- Poor or inadequate project planning
- Inappropriate contract strategy
- Lack of an adequate Project Execution Plan
- Lack of owner resources and leadership
- Poor performance by the project team
- Poor performance by contractors and suppliers
- Lack of skilled resources
- Sales commitments based on unrealistic estimates of project time and cost
- Poor control of changes
- Ineffective materials management—late deliveries of critical items
- Ineffective contract administration
- Ineffective project management
- Inefficient project processes and procedures
- Poor management of new technology
- Turnover of project team

- Ineffective management of technological evolution
- Failure to properly understand user requirements and preferences

- Lack of front-end definition
- Ineffective management of partners
- Poor reservoir management

- Poor design quality causing extensive rework
- Low productivity for engineering or construction
- Poor construction quality
- Ineffective labor relations
- Inadequate permitting planning and execution
- Ineffective community- or public-relations program

- Poor coordination with Operations and Maintenance

- Failure to comply with GMP and related regulations

b. What are the internal risks to project success? (*continued*)

Things to consider:

- Ineffective community- or public-relations program
- Ineffective government-relations program
- Failure to properly anticipate local conditions, regulations, and work practices
- Language difficulties
- Logistics difficulties

Enter the internal project risks on the Risk Mitigation Planning Chart

_____

_____

_____

_____

_____

_____

_____

_____

_____

_____

_____

_____

_____

_____

_____

_____

_____

_____

_____

_____

Internal Risks: The company can control the occurrence of these risks as well as mitigate their impact.

Refer to Issue #

B2. Risk Management Plan

B2: Risk Management Plan

c. What is the probability of each risk occurring?

Things to consider:

- Past experience with each type of risk
- Expected experience on this project
- Surveys of experience of similar projects

Show the probability of each risk on the Risk Mitigation Planning Chart

_____
_____
_____
_____
_____
_____
_____
_____
_____
_____
_____
_____
_____
_____
_____
_____
_____
_____
_____
_____
_____
_____

**Refer to
Issue #**

B2: Risk Management Plan

d. What is the impact of each risk if it should occur?

Things to consider:

- Impact on total project cost
- Impact on overall project schedule
- Impact on profitability
- Impact on overall project difficulty

Show the impact of each risk on the Risk Mitigation Planning Chart

_____
_____
_____
_____
_____
_____
_____
_____
_____
_____
_____
_____
_____
_____
_____
_____
_____
_____
_____
_____
_____

| Refer to |
| Issue # |

B2. Risk Management Plan

B2: Risk Management Plan

e. What risk mitigation steps can be taken to reduce the probability of occurrence and/or the impact?

Things to consider:

- Steps that reduce the probability of occurrence
- Steps that reduce the impact if the risk should occur
- Steps that represent good project management practice
- Steps that may be a departure from common company practice
- Improved project planning
- Contingency planning
- Risk avoidance
- Insurance
- Sharing risks with partners, contractors, vendors

Show the possible Risk Mitigation Steps on the Risk Mitigation Planning Chart

_____
_____
_____
_____
_____
_____
_____
_____
_____
_____
_____
_____
_____
_____
_____
_____
_____

(Note: Risks that have a medium to high probability and a medium to high impact usually must be mitigated; those with lower probabilities or impacts may or may not require mitigation.)

Refer to Issue #

B2: Risk Management Plan

f. What is the relative cost of each risk mitigation step?

Things to consider:

- Cost of resources to provide the added effort
- Cost of premiums to change contract terms
- Cost of incentives

Enter the relative costs of each risk mitigation step on the Risk Mitigation Planning Chart

Refer to
Issue #

B2. Risk Management Plan

B2: Risk Management Plan

g. What is the likelihood that each risk mitigation step will succeed?

Things to consider:

- Experience on similar projects
- Expectations for this project

Enter the likelihood that each risk mitigation step will succeed on the Risk Mitigation Planning Chart

**Refer to
Issue #**

B2: Risk Management Plan

h. Which risk mitigation steps should be undertaken?

Things to consider:

- Probability that the risk will occur
- Impact if the risk occurs
- Relative cost of the mitigation step
- Probability that mitigation will succeed

Indicate ("Yes" or "No") the risk mitigation steps to be taken on the Risk Mitigation Planning Chart

B2. Risk Management Plan

_____

_____

_____

_____

_____

_____

_____

_____

_____

_____

_____

_____

_____

_____

_____

_____

_____

_____

(Note: We will usually take a mitigation step if the probability of occurrence and the impact are high, when the cost of mitigation is low and the probability mitigation will succeed is high. If the probability of occurrence, the probability of success, and/or the impact are low, we may ignore the risk.)

> **Refer to Issue #**

B2: Risk Management Plan

i. Who will be responsible for each risk mitigation step and when should the step be completed?

Things to consider:

- Team members who are best able to implement the mitigation
- Role of contractors, suppliers, and partners
- Timing required to head off risk occurrence

Show the responsibilities and due dates for each risk mitigation step on the Risk Mitigation Planning Chart

_____

_____

_____

_____

_____

_____

_____

_____

_____

_____

_____

_____

_____

_____

_____

_____

_____

_____

_____

_____

Refer to Issue #

B2: Risk Management Plan

j. What are the issues with respect to the Risk Management Plan?

Enter each issue on the Issues List and Path Forward Planning Worksheet.

B3: Organization Plan

Objective: Define the project organization that will meet project objectives and fulfill the scope of work

A project has been said to be nothing more than "people doing work." Project success depends to a great extent on how well these people understand their own role and responsibilities, and those of others.

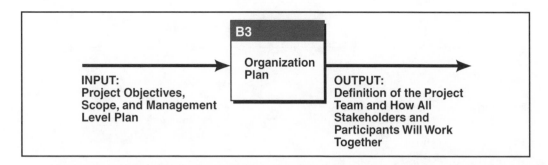

The organization plan is the process by which we develop, document, and communicate how everyone will work together. There are inevitably considerable misunderstandings and issues surrounding this module, so teams should take the time to explore the questions in this section thoroughly.

PMBOK® Guide reference:
2.2 Project Stakeholders
2.3 Organizational Influences
2.4 Key General Management Skills
2.5 Socioeconomic Influences
9.1 Organizational Planning
9.2 Staff Acquisition

Note: Charts for defining roles and responsibilities are provided on the following pages.

Sample LACTI Chart (for Defining Roles and Responsibilities)						
	Person's Name, Title, Function, etc. (be specific)					
Activity Description	Project Mgr.	Project Engr.	Mech. Engr.	Buyer	Elect Engr.	Oper Mgr.
Example:						
Prepare Pump Specification	A	L	T	I	C	C

L = Leads the effort
A = Approves the plan and/or results
C = Consulted for input
T = Tasked to perform the work
I = Informed of plan and/or results, but does not participate in activity

LACTI Chart (for Defining Roles and Responsibilities)						
Description of Activity or Deliverable	**Person's Name, Title, Function, etc. (be specific)**					

L = Leads the effort
A = Approves the plan and/or results
C = Consulted for input
T = Tasked to perform the work
I = Informed of plan and/or results, but does not participate in activity

B3. Organization Plan

LACTI Chart (for Defining Roles and Responsibilities)						
	Person's Name, Title, Function, etc. (be specific)					
Description of Activity or Deliverable						

L = Leads the effort
A = Approves the plan and/or results
C = Consulted for input
T = Tasked to perform the work
I = Informed of plan and/or results, but does not participate in activity

LACTI Chart (for Defining Roles and Responsibilities)						
	Person's Name, Title, Function, etc. (be specific)					
Description of Activity or Deliverable						

L = Leads the effort
A = Approves the plan and/or results
C = Consulted for input
T = Tasked to perform the work
I = Informed of plan and/or results, but does not participate in activity

B3. Organization Plan

B3: Organization Plan

a. What will be the role and responsibilities of the "owner" organization?

Things to consider:

- Setting business goals and project objectives
- Defining and managing the scope of work
- Managing risks
- Arranging project financing and releasing project funds
- Decision-making
- Providing project resources
- Providing project leadership
- Ensuring that the design meets customer requirements
- Developing a Contracting Plan and administering contracts
- Providing owner-furnished materials
- Planning and leading the application of Project Management Best Practices
- Preparing the Project Execution Plan
- Issues management
- Change management
- Interaction with government agencies
- Interaction with the local community
- Interaction with partners
- Interaction with internal and external customers
- Managing project communication interfaces
- Ensuring that a total project integrated master schedule is developed and maintained
- Resource management
- Safety and environmental management

The role and responsibilities of the owner organization are as follows:

_____

_____

_____

_____

_____

_____

_____

(Note: the "owner" organization is the one(s) that will own the asset that the project creates. All other organizations are suppliers or contractors.)

| Refer to |
| Issue # |

B3: Organization Plan

b. Who are the stakeholders who must be represented in the project organization?

Things to consider:

- Other owner/investor organizations
- Lenders
- Marketing
- Product managers
- Regional managers
- Business Unit managers
- Operations and Maintenance
- Research & Development
- End-user customers
- Internal customers
- Suppliers and contractors
- Local community

The stakeholders in this project are:

_____

_____

_____

_____

_____

_____

_____

_____

_____

_____

_____

_____

_____

_____

(Note: "Stakeholders" are project participants who have a functional or financial stake in the outcome.)

Refer to Issue #

B3. Organization Plan

B3: Organization Plan

c. What degree of owner participation is planned? o

Things to consider:

- Owner resources available for project work
- Risks associated with a weak or inadequate level of owner participation and leadership
- Level of owner "due diligence" required for the management of project funds
- Level of project risk (the more risk, the more owner participation is likely to be required)
- Importance of the project to achieving corporate or business-unit strategic goals (the more important, the more owner participation is likely to be required)
- Level of participation necessary to effectively manage work by contractors

- Amount of owner effort required to deal with host-country and local authorities and agencies
- Number and cost of expatriate resources to be assigned to the project location
- Availability of local resources to act on the owner's behalf

The degree of owner participation is to be as follows:

_____

_____

_____

_____

_____

_____

_____

_____

_____

_____

_____

(Note: Research studies confirm that strong, effective owner leadership is a primary factor in determining project success.)

| Refer to |
| Issue # |

B3: Organization Plan

d. How will responsibilities be assigned?

Things to consider:

- Division of work between owner and contractors
- Amount of owner oversight required
- Which organization has the resources needed
- Which organization has the required skill sets
- Where approval authorities will reside

Responsibilities will be assigned as shown in the LACTI chart

_____
_____
_____
_____
_____
_____
_____
_____
_____
_____
_____
_____
_____
_____
_____
_____
_____
_____
_____

Use the "LACTI" chart to develop specific
individual responsibilities (Note: the LACTI
chart (provided at the beginning of this
section) identifies, for each key task or
deliverable, who _Leads_ the work, who
Approves the work, who is _Consulted_ during
the work, who is _Tasked_ to perform the
work, and who is _Informed_ of the results.)

**Refer to
Issue #**

B3. Organization Plan

B3: Organization Plan

e. How will project decisions be made?

Things to consider:

- Grants of authority
- What types of decisions must be made at what levels
- Requirements that decisions be made in time to meet project schedule deadlines
- Level of management review required
- Organizational functions that should be represented in the decision process

Project decisions will be made as follows:

_____

_____

_____

_____

_____

_____

_____

_____

_____

_____

_____

_____

_____

_____

_____

_____

_____

_____

_____

_____

| Refer to |
| Issue # |

B3: Organization Plan

f. What are the critical lines of communication?

Things to consider:

- Within the project team
- Between owner and contractor
- Between the project team and management
- Between the project team and customers
- Language issues
- Time-zone issues
- Distance issues

The critical lines of communication are as follows:

_____

_____

_____

_____

_____

_____

_____

_____

_____

_____

_____

_____

_____

_____

_____

_____

_____

_____

_____

| Refer to |
| Issue # |

B3. Organization Plan

B3: Organization Plan

g. What is the organization chart for the project?

Things to consider:

- Functional organization
- Projectized organization
- Strong matrix organization
- Weak matrix organization
- Owner's project team
- Contractor's project team
- Integrated project team

The project organization chart is as follows:

_____

_____

_____

_____

_____

_____

_____

_____

_____

_____

_____

_____

_____

_____

_____

_____

_____

_____

_____

_____

**Refer to
Issue #**

B3: Organization Plan

h. How will the project team be assured that sufficient resources will be available during each project phase?

Things to consider:

- Internal resources
- Full-time project resources
- Shared project resources
- External sources of resources
- Competition from other internal projects
- Competition from external projects

Sufficient resources for each project phase will be ensured by the following:

_____

_____

_____

_____

_____

_____

_____

_____

_____

_____

_____

_____

_____

_____

_____

_____

_____

_____

_____

_____

Refer to Issue #

B3: Organization Plan

i. What are the issues with respect to the Organization Plan?

**Enter each issue on the Issues List and Path Forward Planning
Worksheet.**

B4: Contract Plan

Objective: Determine how contractors will be utilized to perform the scope of work and meet project objectives

Most major projects require resources that exceed the owner's capacity. Or a project may require skills, knowledge, or technology not available in the owner's organization. For these and many other reasons, most projects involve the use of contractors, i.e., services by firms outside the owner company.

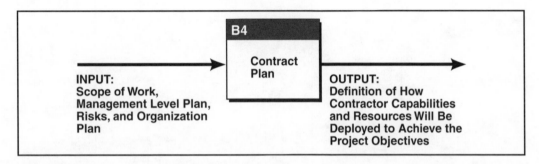

The contract plan defines how contractors will be used on the project—what their scope of work, roles, and responsibilities will be, what resources they will provide, what risks they are expected to take, and how they will be paid for the services they provide. There are many ways to do this, and the contract plan is usually the subject of much debate. Incentives, alliances, and combinations of reimbursable and fixed-price contracts are all possible. There is no one "best" type of contract; the right contract plan for any project will be driven by the strategic project planning results of the previous modules.

PMBOK® Guide reference:
12.1 Procurement Planning
12.2 Solicitation Planning
12.3 Solicitation
12.4 Source Selection

Note: Contract Planning Charts are provided on the following pages.

Contract Planning Chart					
Contract Scope of Work	Contract Type	Potential Bidders	Key Dates	Use Alliance?	Use Incentives?

Contract Planning Chart					
Contract Scope of Work	Contract Type	Potential Bidders	Key Dates	Use Alliance?	Use Incentives?

B4. Contract Plan

Contract Planning Chart					
Contract Scope of Work	Contract Type	Potential Bidders	Key Dates	Use Alliance?	Use Incentives?

B4: Contract Plan

a. What are the responsibilities to be allocated to contractors?

Things to consider:

- Contractor qualifications
- Availability of capable contractors
- Extent of owner participation
- Specialized technology or experience
- Ability of contractors to perform roles traditionally taken by owners
- Use the Work Breakdown Structure to allocate responsibilities
- Domestic content requirements
- Qualifications and capabilities of local contractors
- Experience and capabilities of international contractors in the project location

The responsibilities to be allocated to contractors are as follows:

Refer to
Issue #

B4. Contract Plan

B4: Contract Plan

b. What is the scope of work for each contract?

Things to consider:

- Contractor's capabilities (i.e., avoid assigning a work scope that requires a contractor to do work they are not good at)
- Owner's ability to manage scope
- Owner's ability to manage interfaces between multiple contractors
- Contractor backlog
- Use the Work Breakdown Structure to define contract scope

Enter the scope of work for each contract on the Contract Planning Chart

| Refer to |
| Issue # |

B4: Contract Plan

c. What contracting method will be used for each contract?

Things to consider:

- Market conditions—competitive conditions often favor fixed-price contracting
- Degree of owner control required—if owner requires a lot of control, a reimbursable contract is often preferable
- Allocation of risk—an owner can allocate some risk to a contractor by using a fixed-price contract
- Level of scope definition—a fixed-price contract is seldom appropriate if the scope of work is not well defined and unlikely to change
- Tightness of the schedule—the time required to complete the bidding and award cycle for a fixed-price contract may exceed the time available in a fast-track schedule

Enter the type of contract on the Contract Planning Chart

_____
_____
_____
_____
_____
_____
_____
_____
_____
_____
_____
_____
_____
_____
_____
_____
_____
_____

**Refer to
Issue #**

B4. Contract Plan

B4: Contract Plan

d. Who are the potential bidders for each contract package?

Things to consider:

- Potential bidders with whom owner has experience
- New bidders
- Backlog of potential bidders—can they absorb this additional work?
- Qualifications of the people each bidder would use on the project
- Bidders having desirable proprietary technology
- Bidders having relevant experience
- Alliance contractors
- Local bidders to satisfy domestic content requirements
- Potential bidders with experience at the project location

Enter the potential bidders on the Contract Planning Chart

_____

_____

_____

_____

_____

_____

_____

_____

_____

_____

_____

_____

_____

_____

_____

_____

_____

_____

| Refer to |
| Issue # |

B4: Contract Plan

e. What are the key dates in executing the contract plan?

Things to consider:

- Date on which contractor's scope is fully defined
- Required date of contract award
- Time required to prepare bid packages, screen bidders, have bidders prepare bids, review bids, and award contracts
- Time required for contractor to mobilize and begin work
- Time required to obtain permits

- Time required to obtain government approvals

Enter the key contracting dates on the Contract Planning Chart

_____

_____

_____

_____

_____

_____

_____

_____

_____

_____

_____

_____

_____

_____

_____

_____

_____

Refer to Issue #

B4. Contract Plan

B4: Contract Plan

f. What use will be made of alliance contracts?

Things to consider:

- Existing alliance contracts at the corporate level
- Existing local or business-unit alliances
- Whether the project must use the alliance contractor(s) or is free to solicit other bids
- Forming a new alliance around the current project
- Alliances between contractors and suppliers

Show the planned use of alliances (if any) on the Contract Planning Chart

_____

_____

_____

_____

_____

_____

_____

_____

_____

_____

_____

_____

_____

_____

_____

_____

_____

_____

_____

_____

(Note: alliance contracts are long-term agreements in which the owner commits to using the alliance contractor for certain types of work, in return for which the alliance contractor commits resources to the owner and charges lower rates.)

Refer to Issue #

B4: Contract Plan

g. What use will be made of incentive plans?

Things to consider:

- Incentives based on project cost
- Incentives based on achieving critical milestone dates
- Incentives based on safety performance
- Incentives based on quality performance
- Incentives based on a combination of performance variables
- Bonus-only incentive plans
- Bonus–penalty incentive plans
- Penalty-only plans

Show the use of contract incentives (if any) on the Contract Planning Chart

_____

_____

_____

_____

_____

_____

_____

_____

_____

_____

_____

_____

_____

_____

_____

_____

(Note: The purpose of an incentive plan is to align contractor performance with the owner's project objectives. Incentive plans can address one or more project objectives; they can provide bonuses for good performance and/or penalties for poor performance.)

Refer to Issue #

B4. Contract Plan

B4: Contract Plan

h. What use will be made of existing master agreements?

Things to consider:

- **Master agreements at the corporate level**
- **Master agreements at the project location**

Existing master agreements will be used for the following scopes of work:

_____

_____

_____

_____

_____

_____

_____

_____

_____

_____

_____

_____

_____

_____

_____

_____

_____

_____

_____

_____

_____

_____

(Note: Master agreements are contracts that are already in place, needing only a work order to authorize the contractor to do an assigned scope of work.)

Refer to Issue #

B4: Contract Plan

i. What special requirements, terms, or conditions will be required in the contracts for this project?

Things to consider:

- Project planning and control requirements
- Required use of information systems
- Insurance requirements
- Safety requirements
- Required use of computer-aided design and related tools

- Requirements to comply with local regulations

The following special requirements, terms, and conditions will be reflected in the contracts for this project:

_____

_____

_____

_____

_____

_____

_____

_____

_____

_____

_____

_____

_____

_____

_____

_____

_____

_____

Refer to Issue #

B4. Contract Plan

B4: Contract Plan

j. What special qualifications or capabilities will be required for prequalification of contractors?

Things to consider:

- Specific types of project experience
- Familiarity with the project technology

- Familiarity with the site location

- Experience in the host country
- Ability to work in the language of the host country

The following special qualifications or capabilities will be required for prequalification of contractors for this project:

_____

_____

_____

_____

_____

_____

_____

_____

_____

_____

_____

_____

_____

_____

_____

_____

_____

_____

_____

_____

_____

| Refer to |
| Issue # |

B4: Contract Plan

k. How will the project team be assured that sufficient contractor resources will be available when required?

Things to consider:

- Market demand for contractor services
- Backlog of selected contractors
- Project requirements for skilled personnel with specialized experience
- Contractual commitments to provide certain resources
- Early booking of contractor "space"

The following steps will be taken to ensure that sufficient contractor resources will be available when required:

| Refer to |
| Issue # |

B4: Contract Plan

I. What are the issues with respect to the Contract Plan?

Enter each issue on the Issues List and Path Forward Planning Worksheet.

B5: Best Practices Implementation Plan

Objective: Determine the Project Management Best Practices appropriate to this project, and develop a plan for implementation

Project Management Best Practices, when applied at strategic points in project definition, can have significant benefits in terms of reduced cost, shorter schedule, and higher quality.

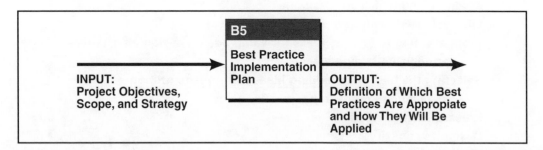

One of the best known Best Practices is value analysis (value engineering). This is a process that can be generally described as optimizing the cost and performance of a design. Since many projects are designed in response to requirements and standards that are unintentionally overstated, significant savings often result. Another Best Practice, for projects that involve construction, is constructability analysis.

PMBOK® Guide reference:
4.3 Overall Change Control

B5: Best Practices Implementation Plan

a. How will Value Analysis principles be applied to this project?

Things to consider:

- Functional analysis (i.e., meeting essential functions at minimum cost)
- Evaluating customer requirements to determine which are essential and which are desirable
- Life-cycle cost analysis
- Brainstorming to identify ideas for value-adding design alternatives
- Evaluating value-adding ideas
- Selecting the best value-adding ideas for implementation
- Process reliability modeling
- Process simplification
- Design to capacity
- Class of plant quality

The plan for the application of value analysis is as follows:

_____

_____

_____

_____

_____

_____

_____

_____

_____

_____

_____

_____

_____

_____

_____

_____

(Note: Value analysis (also called value engineering) is a process for increasing the value of the asset being designed by meeting the right set of design requirements at the lowest life-cycle cost.)

Refer to Issue #

B5: Best Practices Implementation Plan

b. What will be the source of Value Analysis expertise?

Things to consider:

- Engineering contractors
- Internal engineering resources
- Value engineering consultants
- Lessons learned from past projects

The source of value engineering expertise for this project will be as follows:

_____
_____
_____
_____
_____
_____
_____
_____
_____
_____
_____
_____
_____
_____
_____
_____
_____
_____
_____
_____

Refer to Issue #

B5. Best Practices Implementation Plan

B5: Best Practices Implementation Plan

c. At what points in the schedule will value analysis be performed?

Things to consider:

- Early conceptual design (a good time to question requirements)
- Preliminary design (evaluate the initial design, implement alternatives)
- Detailed design (look for design improvements)

Value analysis will be performed at the following points in the schedule:

_____
_____
_____
_____
_____
_____
_____
_____
_____
_____
_____
_____
_____
_____
_____
_____
_____
_____
_____
_____
_____

**Refer to
Issue #**

B5: Best Practices Implementation Plan

d. What aspects of the current design offer opportunities for cost and time savings through value analysis?

Things to consider:

- Design elements whose as-designed cost is more than 300% of the minimum cost to meet essential requirements
- Customer requirements that greatly exceed minimum requirements
- Design elements that represent a large percentage of project cost
- Functional requirements that could be met in a less expensive way
- Use of high-cost materials
- Use of complex designs
- Use of first-of-a-kind technology

The current design offers the opportunity for savings through value analysis in the following areas:

_____

_____

_____

_____

_____

_____

_____

_____

_____

_____

_____

_____

_____

_____

_____

_____

Refer to Issue #

B5: Best Practices Implementation Plan

e. How will constructability principles be applied to this project?

Things to consider:

- Timely application of construction expertise
- Areas of the design and schedule that offer opportunities for cost and time savings through constructability principles

- Maintaining customer service during construction operations
- Phased startup and cutover to new operational systems
- Redundancy of old and new systems
- Minimizing offshore work
- Pre-commissioning
- Location and capabilities of fabrication yards
- Modularization of facilities or subcomponents
- Pre-assembly
- Material selection
- Layout to minimize congestion
- Maximize construction operations at grade
- Construction interference with traffic operations
- Provision of alternative facilities to free areas for construction
- Impact of construction on local neighborhoods

- Plant access for construction forces
- Maximizing work during shutdowns
- Minimizing impact of construction on operations
- Maintaining production operations in accordance with cGMP

- Modularization and pre-assembly in developed nations
- Effective use of local subcontractors

The plan for the application of constructability principles is as follows:

_____

_____

_____

_____

_____

(Note: Constructability is the application of construction expertise to the design and planning process, in order to minimize time, cost, difficulty, and risks of construction.)

Refer to Issue #

B5: Best Practices Implementation Plan

f. What will be the source of constructability expertise?

Things to consider:

- Construction contractors
- Constructability consultants
- Operations management
- Key contractors, subcontractors, and vendors
- Experienced company staff
- Local subcontractors

The sources of constructability expertise will be as follows:

_____

_____

_____

_____

_____

_____

_____

_____

_____

_____

_____

_____

_____

_____

_____

_____

_____

_____

_____

| Refer to |
| Issue # |

B5. Best Practices Implementation Plan

B5: Best Practices Implementation Plan

g. At what points in the schedule will Constructability analysis be performed?

Things to consider:

- Conceptual design and planning—look at overall layout and plans
- Preliminary design and planning—look at plans, elevations, congestion
- Detailed design and planning—look at standards, specifications, designs, detailed plans and schedules, site logistics

Constructability analysis will be performed at the following points in the schedule:

**Refer to
Issue #**

B5: Best Practices Implementation Plan

h. What aspects of the current scope of work and plan offer opportunities for cost and time savings through constructability analysis?

Things to consider:

- Elements of the design that are complex
- Elements of the design that represent difficult working conditions (e.g., above grade, enclosed)
- Elements of the plan that cause congestion of workforce
- Construction operations that are expected to be minimally productive

The elements of the current scope of work and plan that offer opportunities for constructability analysis are as follows:

_____

_____

_____

_____

_____

_____

_____

_____

_____

_____

_____

_____

_____

_____

_____

_____

_____

| Refer to |
| Issue # |

B5. Best Practices
Implementation Plan

B5: Best Practices Implementation Plan

i. How will change management be applied to this project?

Things to consider:

- Responsibilities for scope definition
- Responsibilities for project cost and schedule
- Timely reporting of the status of changes
- Timely estimating of the cost and schedule impact of changes
- Ensuring accountability for the impact of changes on project objectives
- Impact of changes on the contracting plan

Changes will be managed on this project as follows:

_____

_____

_____

_____

_____

_____

_____

_____

_____

_____

_____

_____

_____

_____

_____

_____

_____

_____

_____

_____

_____

| Refer to |
| Issue # |

B5: Best Practices Implementation Plan

j. What other Best Practices are appropriate for this project?

Things to consider:

- 3-D CAD modeling
- Standard design modules

- Process reliability modeling
- Process simplification
- Class of plant quality
- Design to capacity

The other Best Practices that are appropriate for this project are as follows:

_____

_____

_____

_____

_____

_____

_____

_____

_____

_____

_____

_____

_____

_____

_____

_____

_____

_____

**Refer to
Issue #**

141

B5. Best Practices
Implementation Plan

B5: Best Practices Implementation Plan

k. What are the issues with respect to the Best Practice Implementation Plan?

Enter each issue on the Issues List and Path Forward Planning Worksheet.

B6: Team Performance Management Plan

Objective: Ensure a high-performing project team

Most experienced project managers, if hypothetically given total control over one aspect of a project plan, would choose to have a high-performing project team. A high-performing team can make a poorly planned project succeed, just as a poorly performing team can make a well-planned project fail.

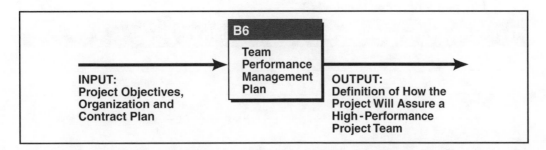

INPUT:
Project Objectives,
Organization and
Contract Plan

B6
Team
Performance
Management
Plan

OUTPUT:
Definition of How the
Project Will Assure a
High-Performance
Project Team

It cannot be taken for granted that a project team will achieve high performance. Top performance requires many things, including ways to measure performance, gain alignment of all participants, ensure effective communication, resolve conflict, and apply lessons learned. None of these is easy, but all are possible with an effective plan.

PMBOK® Guide reference:
2.4 Key General Management Skills
9.3 Team Development

B6. Team Performance
Management Plan

B6: Team Performance Management Plan

a. What will be the method for measuring team effectiveness?

Things to consider:

- Team performance yardsticks
- Metrics for project performance
- Team self-assessment
- Benchmarking by outside parties
- Independent project reviews

Team effectiveness will be measured as follows:

_____

_____

_____

_____

_____

_____

_____

_____

_____

_____

_____

_____

_____

_____

_____

_____

_____

_____

_____

_____

_____

_____

**Refer to
Issue #**

B6: Team Performance Management Plan

b. How will alignment of all stakeholders be ensured?

Things to consider:

- Team alignment workshops
- Strategic Project Planning involving all stakeholders
- Effective communication plan
- Issues management
- Opportunities for stakeholder participation in project planning and decision-making
- Periodic reviews

Alignment of all project stakeholders will be achieved as follows:

_____
_____
_____
_____
_____
_____
_____
_____
_____
_____
_____
_____
_____
_____
_____
_____
_____
_____
_____

Refer to Issue #

B6: Team Performance Management Plan

c. How will conflicts be identified and resolved?

Things to consider:

- Conflict resolution methods
- Allocation of authority for decision-making
- Teambuilding activities
- Conflicts within the project team
- Conflicts between the project team and customers
- Conflicts between the project team and contractors
- Conflicts between contractors

The plan for managing conflicts on this project is as follows:

_____

_____

_____

_____

_____

_____

_____

_____

_____

_____

_____

_____

_____

_____

_____

_____

_____

_____

_____

(Note: Effective teams are evidenced not by lack of conflict but by the effectiveness with which they handle conflict when it occurs.)

Refer to Issue #

B6: Team Performance Management Plan

d. How will trust and cooperation between owner and contractor be ensured?

Things to consider:

- Alignment of commercial motivations
- Sharing common definition of project success
- Commitment of top management
- Contract type and terms that foster trust and cooperation
- Thorough contractor screening process
- Owner leadership
- Joint owner–contractor strategic planning sessions
- Clear definition of specific owner and contractor responsibilities
- Effective issues management
- Incentives that align priorities

Owner–contractor trust and cooperation will be ensured by:

_____

_____

_____

_____

_____

_____

_____

_____

_____

_____

_____

_____

_____

_____

_____

_____

_____

_____

**Refer to
Issue #**

B6. Team Performance
Management Plan

B6: Team Performance Management Plan

e. How will the contractor's and owner's goals be aligned?

Things to consider:

- Incentive plans
- Recognition programs
- Fair and effective contractor performance metrics based on owner's goals
- Owner–contractor goal-setting workshops

Alignment of owner's and contractor goals will be accomplished as follows:

_____

_____

_____

_____

_____

_____

_____

_____

_____

_____

_____

_____

_____

_____

_____

_____

_____

_____

_____

_____

**Refer to
Issue #**

B6: Team Performance Management Plan

f. How will effective project leadership be ensured?

Things to consider:

- Clear definition of goals, objectives, and priorities
- Effective planning
- Timely and effective decision-making
- Open communication
- Plans, decisions, and actions that are always consistent with project goals, objectives, and priorities
- Selection of project managers who have strong leadership skills

Effective project leadership will be ensured as follows:

_____

_____

_____

_____

_____

_____

_____

_____

_____

_____

_____

_____

_____

_____

_____

_____

_____

_____

(Note: Project leadership—quite different from "project management"—is an owner responsibility that involves defining, communicating, and getting buy-in to a vision of success, building trust through consistency, and effectively performing the owner's project management responsibilities.)

Refer to Issue #

B6. Team Performance Management Plan

B6: Team Performance Management Plan

g. What are the relevant lessons learned from other projects that can be used to improve this project's performance?

Things to consider:

- Positive lessons worth repeating
- Negative lessons to be avoided
- Technical lessons learned
- Management lessons learned

The relevant lessons learned from other projects are as follows:

_____

_____

_____

_____

_____

_____

_____

_____

_____

_____

_____

_____

_____

_____

_____

_____

_____

_____

_____

_____

_____

_____

Refer to Issue #

B6: Team Performance Management Plan

h. What will be the method for capturing lessons learned from each phase of this project?

Things to consider:

- Lessons-learned workshops
- Post-project appraisals
- Lessons learned log or database
- Lessons learned by contractors
- Lessons learned by owner

The method for capturing lessons learned on this project will be as follows:

_____

_____

_____

_____

_____

_____

_____

_____

_____

_____

_____

_____

_____

_____

_____

_____

_____

_____

_____

_____

**Refer to
Issue #**

B6. Team Performance
Management Plan

B6: Team Performance Management Plan

i. What methods or events are planned to ensure team effectiveness?

Things to consider:

- Teambuilding activities
- Team alignment workshops
- Psychological profiles of key team members
- Team effectiveness training
- Team effectiveness metrics

The following methods or events will be implemented to ensure team effectiveness:

_____

_____

_____

_____

_____

_____

_____

_____

_____

_____

_____

_____

_____

_____

_____

_____

_____

_____

_____

_____

_____

Refer to Issue #

B6: Team Performance Management Plan

j. What are the issues with respect to Team Performance?

Enter each issue on the Issues List and Path Forward Planning Worksheet.

B6. Team Performance Management Plan

VI. *The Strategic Project Planner*

Part C: Defining the Tools for Success

As the project progresses, the focus of strategic planning becomes planning how the key project performance variables will be controlled so that excellent performance is achieved.

For most projects, the key project performance variables are (not in order of priority):

- Time
- Cost
- Quality
- Safety and environmental compliance

In addition, materials will have to be managed during project execution, and effective communications ensured.

The resulting process for Defining the Tools for Success is shown below.

C. Defining the Tools for Success

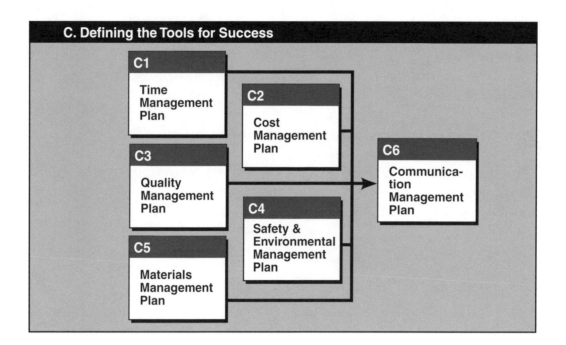

C1: Time Management Plan

Objective: Determine how schedules will be developed, and progress controlled

Project scope of work can be expressed as the activities that must be performed in order to create the expected project deliverables. These activities require both time and resources, and this module addresses how those will be managed.

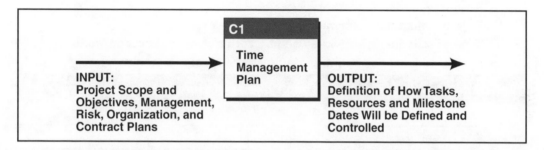

A key element of time management is the development, maintenance, and application of a resource-loaded, critical path schedule. Many projects fail to give adequate attention to scheduling both tasks and resources, in spite of the powerful software available today. An effective Time Management Plan is the key to meeting the project's schedule objectives.

PMBOK® Guide reference:
4.3 Overall Change Control
5.5 Scope Change Control
6.1 Activity Definition
6.2 Activity Sequencing
6.3 Activity Duration Estimating
6.4 Schedule Development
6.5 Schedule Control
7.1 Resource Planning
10.3 Performance Reporting

C1: Time Management Plan

a. How will schedules be prepared during each phase of the project?

Things to consider:

- Milestone schedules in the early stages of project definition
- Critical path schedules as the project is better defined
- Integrated Total Project Master Schedule, including owner and contractor activities for all project phases
- Resource-loaded schedules
- Type of scheduling software used by the owner
- Type of scheduling software used by contractors and subcontractors
- How schedules will be integrated into the Master Schedule

Schedules will be prepared at each phase of the project as follows:

_____

_____

_____

_____

_____

_____

_____

_____

_____

_____

_____

_____

_____

_____

_____

_____

_____

_____

_____

_____

_____

| Refer to |
| Issue # |

C1: Time Management Plan

b. How will the schedule impact of changes be estimated and approved?

Things to consider:

- Impact of changes on the critical path
- Impact of changes on resource requirements
- Impact of changes on the forecast time and cost to complete
- Change-management procedures
- Methods for minimizing time to prepare change estimates
- Accuracy required for change estimates
- How the schedule impact of changes will be reflected in contract administration
- Allocation of responsibilities for defining and estimating the schedule impact of changes
- Change reports, tracking, and trending

The schedule impact of changes will be estimated and approved as follows:

_____

_____

_____

_____

_____

_____

_____

_____

_____

_____

_____

_____

_____

_____

_____

_____

_____

(Note: Virtually all changes have some impact on the task-and-resource schedule.)

Refer to Issue #

159

C1. Time Management Plan

C1: Time Management Plan

c. How will the Integrated Project Master Schedule (IPMS) be prepared and maintained?

Things to consider:

- Owner vs. contractor responsibility for the IPMS preparation and updating
- How subproject schedules will be integrated
- How the IPMS will be maintained
- How updates to the IPMS will be reflected back into subproject schedules
- How overall resource requirements will be analyzed

The plan for preparing and maintaining an Integrated Project Master Schedule is as follow:

_____

_____

_____

_____

_____

_____

_____

_____

_____

_____

_____

_____

_____

_____

(Note: The Integrated Project Master Schedule (IPMS) shows all tasks, dependencies, dates, and responsibilities for all the current and future project phases. It shows tasks for the owner (including project and startup/operations tasks), contractors, and key suppliers. It is often prepared and maintained by merging subproject schedules prepared by the various parties to the project.)

Refer to Issue #

C1: Time Management Plan

d. How will skilled human resources be scheduled during each phase of the project?

Things to consider:

- Owner resources
- Contractor resources
- Resource-loaded schedules
- Use of resource scheduling software
- Resource availability
- Specialized skill requirements
- Resource density
- Project resources
- Operations resources

Skilled resources will be scheduled as follows:

Refer to Issue #

C1: Time Management Plan

e. How will competition from other projects—both internal and external—impact resource availability?

Things to consider:

- Other owner projects in same timeframe
- Other contractor projects in same timeframe
- Projects in same area
- Projects requiring the same skilled resources
- Projects requiring specialty equipment

The impact of other projects on the availability of resources is as follows:

_____

_____

_____

_____

_____

_____

_____

_____

_____

_____

_____

_____

_____

_____

_____

_____

_____

_____

_____

_____

| Refer to
| Issue # |

C1: Time Management Plan

f. How does the current assessment of resource availability compare with known or anticipated project requirements?

Things to consider:

- In-house resource availability
- Supplemental resources from contractors
- Transfer of resources from other projects
- Changes to the contract strategy
- Impact on schedule of resource shortfalls

The current assessment of resource availability is as follows:

_____

_____

_____

_____

_____

_____

_____

_____

_____

_____

_____

_____

_____

_____

_____

_____

_____

_____

_____

Refer to Issue #

C1. Time Management Plan

C1: Time Management Plan

g. How will progress be measured and controlled?

Things to consider:

- Use of earned value techniques
- Analyses of physical (earned) progress vs. percent time, money, and effort spent
- Use of pre-existing contractor yardsticks
- Tying of progress measurement to progress payments
- Contractor vs. owner role in measuring progress
- How progress of owner activities will be measured
- Software applications to record progress and calculate cost and schedule performance statistics

Progress will be measured and controlled as follows:

_____
_____
_____
_____
_____
_____
_____
_____
_____
_____
_____
_____
_____
_____
_____
_____
_____

Note: Earned value is a method for measuring progress and performance that compares the amount of physical work achieved with the time, effort, and money spent.

Refer to Issue #

C1: Time Management Plan

h. How will schedule variances (planned vs. achieved) be identified and corrected?

Things to consider:

- Preparation and maintenance of a baseline schedule
- Updating of the baseline schedule for approved changes
- Defining actual vs. baseline variances
- Defining forecast final date vs. baseline schedule
- How the root cause of a schedule variance will be defined and corrective actions determined
- How the variance will be used to forecast time and cost to complete
- Responsibility for defining and reporting variances
- Responsibility for taking corrective action to correct variances

Schedule variances will be defined and corrected as follows:

| Refer to |
| Issue # |

C1: Time Management Plan

i. How will the responsibilities for planning and scheduling be allocated?

Things to consider:

- Owner responsibilities—the owner is usually responsible for the overall project planning. This responsibility is discharged by:
  - ➢ setting schedule objectives
  - ➢ developing milestone schedules
  - ➢ scheduling owner activities
  - ➢ maintaining the Integrated Project Master Schedule
  - ➢ reviewing contractor schedules
  - ➢ ensuring that a current schedule baseline for the total project is maintained to reflect approved changes
- Contractor responsibilities—the contractor is usually responsible for planning and scheduling the scope of work in the contract and may also have scheduling responsibilities for other aspects of the project. This responsibility is discharged by:
  - ➢ developing detailed, resource-loaded critical path schedules of contractor and subcontractor activities
  - ➢ setting baseline schedules and updating for changes
  - ➢ managing subcontractor and supplier schedules

The responsibilities for planning and scheduling will be allocated as follows:

_____

_____

_____

_____

_____

_____

_____

_____

_____

_____

_____

| Refer to |
| Issue # |

C1: Time Management Plan

j. How will the responsibilities for progress reporting and control be allocated?

Things to consider:

- Owner responsibilities typically include:
  - ➤ defining progress measurement and reporting expectations
  - ➤ reviewing progress reports and analyzing schedule variances
  - ➤ taking corrective actions as necessary
  - ➤ developing independent forecasts of time to complete the total project
  - ➤ administration of prime contracts
- Contractor responsibilities typically include:
  - ➤ measuring physical progress for contract and subcontractor scope of work
  - ➤ calculating schedule productivity (i.e., Schedule Performance Index)
  - ➤ preparing timely and unbiased progress reports
  - ➤ identifying and analyzing schedule variances
  - ➤ defining and executing corrective actions
  - ➤ forecasting time to complete the contractor's scope
  - ➤ administration of subcontracts

The responsibilities for progress measurement and control will be allocated as follow:

_____

_____

_____

_____

_____

_____

_____

_____

_____

_____

Refer to Issue #

C1. Time Management Plan

C1: Time Management Plan

k. How will the time and resources needed to complete the project be forecast?

Things to consider:

- Forecasts based on trend analysis
- Forecasts based on earned value statistics (e.g., Schedule Performance Index)
- Forecasts based on detailed schedule updates reflecting the remaining scope of work, tasks, resources, and duration
- Forecasts based on historical data from similar projects
- Forecasts reflecting resource availability for the remaining work
- Use of tracking curves for graphical forecasting
- Periodic use of independent forecasts as a "reality check"

The time and resources to complete the project will be forecast as follows:

_____

_____

_____

_____

_____

_____

_____

_____

_____

_____

_____

_____

_____

_____

_____

_____

_____

| Refer to |
| Issue # |

C1: Time Management Plan

I. How will schedule contingency be administered?

Things to consider:

- Use of contingency "buffers"
- Use of schedule contingency in major activities
- Use of target schedules (with no contingency) that are adjusted for approved changes
- Allocation of schedule contingency between owner and contractor
- Re-forecasting (and reducing) contingency as the work progresses to reflect reduced uncertainty

Schedule contingency will be administered as follows:

_____

_____

_____

_____

_____

_____

_____

_____

_____

_____

_____

_____

_____

_____

_____

_____

_____

_____

_____

_____

_____

Refer to Issue #

C1: Time Management Plan

m. What are the issues with respect to time management?

Enter each issue on the Issues List and Path Forward Planning Worksheet.

C2: Cost Management Plan

Objective: Determine how costs will be estimated and controlled

Project economics are often sensitive to project cost, and many projects are driven by the need to stay under budget. Cost management is often a considerable challenge, since everything that happens on a project is likely to have some impact on cost.

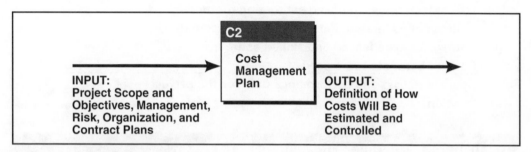

Cost and time management are closely related as both are driven by the scope of work, resources, and productivity trends. Effective cost management requires a sound and complete cost estimate to serve as a baseline, and timely cost tracking to determine variances.

PMBOK® Guide reference:
4.3 Overall Change Control
5.5 Scope Change Control
7.1 Resource Planning
7.2 Cost Estimating
7.3 Cost Budgeting
7.4 Cost Control
10.3 Performance Reporting

C2. Cost Management Plan

C2: Cost Management Plan

a. How will cost estimates be prepared to support funding decisions?

Things to consider:

- Timing of estimates
- Degree of definition of project scope and plans at the time the estimates are prepared
- Use of factored vs. detailed estimating methods
- Sources of estimating data (internal vs. external)
- Sources of experienced and skilled estimators
- Code of accounts to be used

- Sources of cost and performance data for the project location

The plan for preparing cost estimates is as follows:

_____

_____

_____

_____

_____

_____

_____

_____

_____

_____

_____

_____

_____

_____

_____

_____

_____

_____

| Refer to |
| Issue # |

C2: Cost Management Plan

b. How will the cost of changes be estimated and approved?

Things to consider:

- Change in management procedures
- Methods for minimizing time required to prepare change estimates
- Accuracy required for change estimates
- How the cost of changes will be reflected in contract administration
- Allocation of responsibilities for defining and estimating changes
- Change reports, tracking, and trending

The plan for estimating the cost of changes is as follows:

_____

_____

_____

_____

_____

_____

_____

_____

_____

_____

_____

_____

_____

_____

_____

_____

_____

_____

_____

_____

| Refer to |
| Issue # |

173

C2. Cost Management Plan

C2: Cost Management Plan

c. How will scope and other changes to the project be controlled?

Things to consider:

- Scope changes—usually require management approval for supplemental funding
- Design changes—usually to optimize the design
- Field changes—usually to facilitate construction
- Startup changes—usually to facilitate startup
- Potential changes—changes being considered
- Pending changes—changes being evaluated
- Approved changes—changes authorized for implementation
- Contingency rundown to offset the cost of approved changes
- Reporting, tracking, analysis, trending, and forecasting of changes
- Who approves changes?

Scope and other changes to the project will be controlled as follows:

_____

_____

_____

_____

_____

_____

_____

_____

_____

_____

_____

_____

_____

_____

_____

_____

_____

_____

_____

| Refer to |
| Issue # |

C2: Cost Management Plan

d. How will cost tracking and performance analysis be accomplished?

Things to consider:

- Capture and reporting of actual costs
- Owner and contractor codes of accounts
- Application of earned value methods for determining cost performance
- Definition of variances between actual and budget
- Analysis of variances to determine the cause
- Cost reporting

Project costs will be tracked as follows:

_____

_____

_____

_____

_____

_____

_____

_____

_____

_____

_____

_____

_____

_____

_____

_____

_____

_____

_____

Refer to Issue #

C2: Cost Management Plan

e. How will cost variances (spent vs. budget) be identified and corrected?

Things to consider:

- Preparation and maintenance of a current control estimate
- Defining variances between actual costs and current control estimate
- Defining forecast final cost vs. budget
- How the root cause of a cost variance will be defined, and corrective actions determined
- How the variance will be used to forecast time and cost to complete
- Responsibility for defining and reporting variances
- Responsibility for taking corrective action to correct variances

Cost variances will be identified and corrected as follows:

| Refer to |
| Issue # |

C2: Cost Management Plan

f. What will be the interface between project cost control and the in-house accounting systems?

Things to consider:

- Capture of actual costs
- Reconciliation of project costs to accounting records
- Integration of project cost management system with company accounting systems
- Interfaces between owner and contractor cost management systems

The interface between project and company cost management systems will be managed as follows:

_____

_____

_____

_____

_____

_____

_____

_____

_____

_____

_____

_____

_____

_____

_____

_____

_____

_____

_____

_____

| Refer to |
| Issue # |

C2. Cost Management Plan

C2: Cost Management Plan

g. How will the responsibilities for cost estimating be allocated?

Things to consider:

- Owner responsibilities—the owner is usually responsible for the overall project cost. This responsibility is discharged by:
  - ➤ setting cost objectives
  - ➤ developing conceptual and scoping estimates
  - ➤ estimating the cost of owner's activities and materials
  - ➤ reconciling current cost estimates with previous estimates and with the costs for similar projects
  - ➤ establishing the scope, planning, and estimating basis for contractor-prepared estimates
  - ➤ reviewing contractor estimates
  - ➤ ensuring that a current control estimate for the total project is maintained to reflect approved changes
  - ➤ capturing actual cost data for use in preparing future estimates
- Contractor responsibilities—the contractor is usually responsible for estimating the scope of work in the contract and may also have specific estimating responsibilities for other aspects of the project. This responsibility is discharged by:
  - ➤ estimating the cost of performing the contractor's scope of work
  - ➤ setting up and maintaining the current control estimate to reflect approved changes
  - ➤ directing the preparation of cost estimates from subcontractors
  - ➤ reconciling actual costs to the current cost estimate

The allocation of responsibilities for preparing cost estimates is as follows:

_____

_____

_____

_____

_____

_____

_____

_____

_____

| Refer to |
| Issue # |

C2: Cost Management Plan

h. How will the responsibilities for cost reporting and control be allocated?

Things to consider:

- Owner responsibilities typically include:
  - ➢ defining cost performance measurement and reporting expectations
  - ➢ reviewing cost reports and analyzing cost variances
  - ➢ taking corrective actions as necessary
  - ➢ providing independent forecasts of cost to complete the total project
  - ➢ administration of prime contracts
- Contractor responsibilities typically include:
  - ➢ measuring cost productivity for contract and subcontractor scope of work (i.e., Cost Performance Index)
  - ➢ tracking expended and committed costs, and value of work done
  - ➢ preparing timely and unbiased cost reports
  - ➢ identifying and analyzing cost variances
  - ➢ defining and executing corrective actions
  - ➢ forecasting cost to complete the contractor's scope
  - ➢ providing final closeout reports
  - ➢ administration of subcontracts

The responsibilities for cost reporting and control will be allocated as follows:

Refer to
Issue #

C2: Cost Management Plan

i. How will the cost to complete the project be forecast?

Things to consider:

- Forecasts based on trend analysis
- Forecasts based on earned value statistics (e.g., Cost Performance Index)
- Forecasts based on detailed schedule updates reflecting the remaining scope of work, tasks, resources, costs, and durations
- Forecasts based on historical data from similar projects
- Forecasts reflecting cost projections for outstanding purchase orders and contracts
- Use of tracking curves for graphical forecasting
- Periodic use of independent forecasts as a "reality check"

Cost to complete will be forecast as follows:

_____
_____
_____
_____
_____
_____
_____
_____
_____
_____
_____
_____
_____
_____
_____
_____
_____
_____
_____

> **Refer to**
> **Issue #**

C2: Cost Management Plan

j. How will cost contingency be administered?

Things to consider:

- Use of a contingency rundown curve to set contingency in the current control estimate
- Periodic re-forecasting (and reduction) of contingency as the work progresses and uncertainties are reduced
- Allocation of contingency to major cost accounts
- Allocation of cost contingency between owner and contractor
- Use of target control estimate (with no contingency) that is adjusted for approved changes
- How and when to "give back" contingency funds when the project is underrunning

Cost contingency will be administered as follows:

Refer to
Issue #

C2. Cost Management Plan

C2: Cost Management Plan

k. What are the issues with respect to cost management?

Enter each issue on the Issues List and Path Forward Planning Worksheet.

C3: Quality Management Plan

Objective: Determine how quality will be defined and controlled

Many project decisions involve the tradeoff among cost, schedule, and quality. Quality, usually defined as "conformance to customer requirements," can mean different things in different projects. For example, one project may focus on the quality of work done by a contractor—perhaps measured in terms of errors and rework—while another may focus on the quality of the project deliverable—perhaps measured by customer satisfaction.

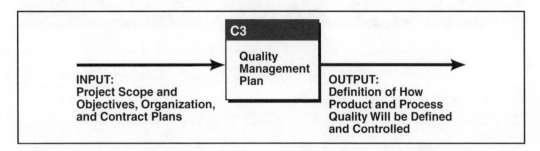

It is important that the Quality Management Plan clarify how the project will define and control quality, and how responsibilities will be allocated.

PMBOK® Guide reference:
8.1 Quality Planning
8.2 Quality Assurance
8.3 Quality Control

C3: Quality Management Plan

a. How will quality be defined for the project?

Things to consider:

- Conformance to customer requirements
- Fitness to purpose
- Amount of rework
- Certification by third parties
- Adherence to specifications
- Quality of project deliverables

The project's definition of quality is as follows:

_____

_____

_____

_____

_____

_____

_____

_____

_____

_____

_____

_____

_____

_____

_____

_____

_____

_____

_____

_____

_____

Refer to Issue #

C3: Quality Management Plan

b. How will responsibilities for quality control be allocated between the project team and contractors?

Things to consider:

- Owner responsibilities typically include:
  - ➤ setting quality standards and expectations
  - ➤ reviewing the contractor's Quality Management Plan
  - ➤ performing quality audits
  - ➤ witnessing tests by contractors and key suppliers of owner-supplied materials and equipment
  - ➤ taking corrective actions to assure the effectiveness of contractor's quality control activities
- Contractor responsibilities typically include:
  - ➤ developing the Quality Management Plan
  - ➤ performing quality control activities
  - ➤ witnessing tests by vendors and subcontractors
  - ➤ performing inspections
  - ➤ developing and documenting work processes
  - ➤ taking corrective actions to maintain quality

Responsibilities for quality control will be allocated as follows:

Refer to
Issue #

C3. Quality Management Plan

C3: Quality Management Plan

c. What methods will be used to define and control quality?

Things to consider:

- Quality Management Plans
- Quality tracking and analysis
- Sampling
- Root Cause Analysis

Responsibilities for quality control will be allocated as follows:

_____

_____

_____

_____

_____

_____

_____

_____

_____

_____

_____

_____

_____

_____

_____

_____

_____

_____

_____

_____

_____

_____

| Refer to |
| Issue # |

C3: Quality Management Plan

d. What are the issues with respect to Quality Management?

Enter each issue on the Issues List and Path Forward Planning Worksheet.

C3. Quality Management Plan

C4: Safety and Environmental Management Plan

Objective: Determine how safety and environmental objectives will be achieved

Many projects, such as the construction of manufacturing or transportation facilities, have a primary focus on safety and environmental management. For these projects, people must be protected while the project is progressing as well as when they are operating or maintaining the facilities delivered by the project. In addition, environmental regulations must be complied with at all times and in all project phases.

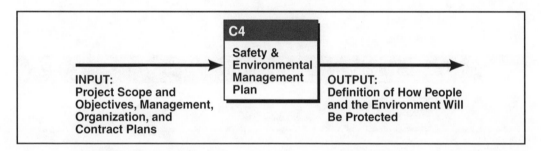

Failure to plan how safety objectives and environmental statutory requirements will be met can easily cause a project substantial delays and extra costs—not to mention the ethical issues associated with poor performance in these areas. An effective Safety and Environmental Management Plan will reduce the risk of problems in this area and improve project performance.

PMBOK® Guide reference:
11.1 Risk Identification
11.2 Risk Quantification
11.3 Risk Response Development
11.4 Risk Response Control

C4. Safety and Environmental Management Plan

C4: Safety and Environmental Management Plan

a. What are the project's safety objectives?

Things to consider:

- Operational safety
- Product safety

- Construction safety

- Safety of data

The project's safety objectives are as follows:

_____

_____

_____

_____

_____

_____

_____

_____

_____

_____

_____

_____

_____

_____

_____

_____

_____

_____

_____

_____

| Refer to |
| Issue # |

C4: Safety and Environmental Management Plan

b. What are the requirements for security and how will they be met?

Things to consider:

- Security of the asset
- Security of supplies and inventory
- Security of people
- Security of information
- Security of construction equipment and materials

- Special location-related security concerns

The requirements for security will be met as follows:

_____

_____

_____

_____

_____

_____

_____

_____

_____

_____

_____

_____

_____

_____

_____

_____

_____

Refer to Issue #

C4: Safety and Environmental Management Plan

c. How will responsibilities for safety performance be allocated between the project management team and contractors?

Things to consider:

- Owner responsibilities typically include:
  - ➤ setting safety standards and expectations
  - ➤ preparing Safety Management Plans for owner activities
  - ➤ reviewing contractor safety performance, plans, practices, and procedures
  - ➤ providing safety training for owner's staff
  - ➤ monitoring safety performance
  - ➤ taking timely and effective corrective actions as necessary.
- Contractor responsibilities typically include:
  - ➤ preparing Safety Management Plans and detailed procedures
  - ➤ ensuring strict compliance with all safety requirements and procedures
  - ➤ taking timely and effective corrective actions as necessary
  - ➤ providing safety training for contractor's staff
  - ➤ providing safety equipment and resources

The responsibilities for safety performance will be allocated as follows:

_____

_____

_____

_____

_____

_____

_____

_____

_____

_____

_____

_____

Refer to Issue #

C4: Safety and Environmental Management Plan

d. How will safety performance be measured, evaluated, and continuously improved?

Things to consider:

- Safety performance statistics
- Safety audits
- Safety incentive plans

Safety performance will be measured and continuously improved as follows:

_____

_____

_____

_____

_____

_____

_____

_____

_____

_____

_____

_____

_____

_____

_____

_____

_____

_____

_____

_____

**Refer to
Issue #**

C4: Safety and Environmental Management Plan

e. What environmental regulations apply to the project?

Things to consider:

- Air emissions
- Solid and liquid waste
- Hazardous waste
- Permits
- Impact on wildlife
- Wetlands
- Federal regulations
- State regulations
- Local regulations
- Host-country regulations
- Local regulations

The following environmental regulations apply to this project:

_____
_____
_____
_____
_____
_____
_____
_____
_____
_____
_____
_____
_____
_____
_____
_____
_____

Refer to
Issue #

C4: Safety and Environmental Management Plan

f. What is the plan for ensuring compliance with all environmental regulations?

Things to consider:

- Permitting plan
- Product specifications

- Protection from spills

- Requirements for construction
- Soil remediation
- Disposal of hazardous materials
- Process design
- Design and capacity of emissions clean-up equipment

- Plan for meeting host-country requirements

The plan for ensuring compliance with environmental regulations is as follows:

_____

_____

_____

_____

_____

_____

_____

_____

_____

_____

_____

Refer to Issue #

C4: Safety and Environmental Management Plan

g. What changes in statutory environmental compliance requirements are anticipated during the project?

Things to consider:

- Government changes
- Evolutionary changes to regulations
- Changes in "best available technology"
- Changes in company policy

The following changes in statutory environmental requirements are anticipated:

_____

_____

_____

_____

_____

_____

_____

_____

_____

_____

_____

_____

_____

_____

_____

_____

_____

_____

_____

_____

_____

_____

Refer to Issue #

C4: Safety and Environmental Management Plan

h. What permits will be required?

Things to consider:

- Building permits
- Occupancy permits

- Operating permits

The following permits will be required:

_____

_____

_____

_____

_____

_____

_____

_____

_____

_____

_____

_____

_____

_____

_____

_____

_____

_____

_____

_____

_____

Refer to Issue #

C4: Safety and Environmental Management Plan

i. How will responsibilities for environmental compliance be allocated between the project team and contractors?

Things to consider:

- Owner responsibilities typically include:

  - ➤ ensuring that all statutory requirements are known and understood
  - ➤ ensuring that an Environmental Management Plan is in place with appropriate procedures
  - ➤ ensuring that contractors perform in accordance with all requirements
  - ➤ conducting periodic audits for compliance
  - ➤ performing emergency response activities when needed
  - ➤ maintaining open and effective communication with government agencies
- Contractor responsibilities typically include:
  - ➤ arranging for necessary permits
  - ➤ ensuring that all work is done in full compliance
  - ➤ providing necessary environmental protection facilities

The responsibilities for environmental compliance will be allocated as follows:

_____

_____

_____

_____

_____

_____

_____

_____

_____

_____

_____

_____

| Refer to |
| Issue # |

C4: Safety and Environmental Management Plan

j. What are the issues with respect to safety and environmental management?

Enter each issue on the Issues List and Path Forward Planning Worksheet.

C4. Safety and Environmental Management Plan

C5: Materials Management Plan

Objective: Determine how material resources will be secured and deployed

Although much attention is given to the planning and managing of human resources, it is also self-evident that these resources cannot be effective without the timely supply of the materials they need to work with.

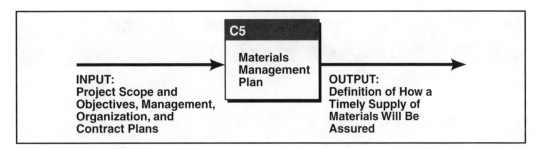

The Materials Management Plan ensures that long-lead items are ordered in a timely fashion, that critical path materials are available when needed, and that responsibilities for every step in the materials management process are clearly defined.

PMBOK® Guide reference:
12.1 Procurement Planning
12.2 Solicitation Planning
12.3 Solicitation
12.4 Source Selection
12.5 Contract Administration
12.6 Contract Closeout

C5: Materials Management Plan

a. How will timely supply of critical, long-lead materials be ensured?

Things to consider:

- Early purchasing of long-lead items
- Owner purchasing vs. contractor purchasing
- Premiums for on-time delivery
- Expediting

Timely supply of long-lead and critical items will be ensured as follows:

_____
_____
_____
_____
_____
_____
_____
_____
_____
_____
_____
_____
_____
_____
_____
_____
_____
_____
_____
_____
_____
_____

Refer to Issue #

C5: Materials Management Plan

b. How will storage and inventory of project materials be managed?

Things to consider:

- Just-in-time deliveries
- On-site storage
- Rented storage space
- Warehousing procedures
- Allocation of storage and inventory responsibilities
- Barcoding
- Location and size of staging areas

- Security of storage areas

Storage and inventory of project materials will be managed as follows:

_____

_____

_____

_____

_____

_____

_____

_____

_____

_____

_____

_____

_____

_____

_____

_____

_____

_____

**Refer to
Issue #**

C5: Materials Management Plan

c. What use of material supply alliances is anticipated?

Things to consider:

- Alliances for the supply of critical equipment
- Alliances for the supply of bulk materials
- Owner's alliances with suppliers
- Contractor's alliances with suppliers

Material supply alliances are anticipated to be used as follows:

_____

_____

_____

_____

_____

_____

_____

_____

_____

_____

_____

_____

_____

_____

_____

_____

_____

_____

_____

_____

_____

**Refer to
Issue #**

C5: Materials Management Plan

d. What existing policies or agreements impact the selection of vendors?

Things to consider:

- Preferred vendor lists
- Existing alliances
- Supplier qualification requirements
- Community relations
- Domestic content

The policies or agreements that impact the selection of vendors is as follows:

_____

_____

_____

_____

_____

_____

_____

_____

_____

_____

_____

_____

_____

_____

_____

_____

_____

_____

_____

| Refer to |
| Issue # |

C5: Materials Management Plan

e. How will timely provision of vendor design information be ensured?

Things to consider:

- Electronic transfer
- Detailed scheduling of information requirements
- Identification of critical information requirements

Timely provision of vendor design information will be ensured as follows:

_____

_____

_____

_____

_____

_____

_____

_____

_____

_____

_____

_____

_____

_____

_____

_____

_____

_____

_____

_____

_____

_____

_____

Refer to Issue #

C5: Materials Management Plan

f. What are the issues with respect to materials management?

Enter each issue on the Issues List and Path Forward Planning Worksheet.

C6: Communication Management Plan

Objective: Determine how stakeholders, the project team, contractors, and suppliers will provide, receive, and utilize project information

One of the greatest challenges for a project manager is the assurance of effective communication. A project has many communication interfaces and an overwhelming amount of data and information to manage. The Communications Management Plan describes how information systems and communications technology will be used to get the right information to the right people at the right time.

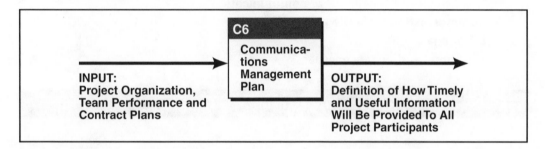

PMBOK® Guide reference:
10.1 Communications Planning
10.2 Information Distribution
10.3 Performance Reporting
10.4 Administrative Closure

C6: Communication Management Plan

a. How will effective communication be ensured?

Things to consider:

- Communication between members of the core project team
- Communication between the core and extended project teams
- Communication between the project team and management
- Communication between the project team and customers
- Communication between the project team and contractors/suppliers
- Communication between stakeholders
- Need to ensure timeliness of communication
- Communication technology
- Language issues
- Time zone and distance issues

Effective communication will be ensured as follows:

_____

_____

_____

_____

_____

_____

_____

_____

_____

_____

_____

_____

_____

_____

_____

_____

_____

_____

_____

| Refer to |
| Issue # |

C6: Communication Management Plan

b. What will be the application of project management software for communication on this project?

Things to consider:

- Planning and scheduling
- Resource management
- Document management
- Cost estimating
- Cost reporting and control
- Progress and performance management
- Change management
- Progress reporting
- Cost and schedule forecasting
- Project photos and videos

Project management software will be applied on this project as follows:

_____

_____

_____

_____

_____

_____

_____

_____

_____

_____

_____

_____

_____

_____

_____

_____

_____

_____

| Refer to
| Issue # |

C6. Communication Management Plan

C6: Communication Management Plan

c. What will be the application of Web-based project and information management technology?

Things to consider:

- Communication between various worksites
- "Virtual war room"
- Instant updating of information
- Central storehouse of information—accessible to all

The project will use a Web-based project and information management technology as follows:

_____

_____

_____

_____

_____

_____

_____

_____

_____

_____

_____

_____

_____

_____

_____

_____

_____

_____

_____

_____

_____

_____

| Refer to |
| Issue # |

C6: Communication Management Plan

d. How will information needs be identified and documented?

Things to consider:

- Information required by the owner's core PM team
- Information required by the owner's extended PM team, e.g.,
  - ➤ Purchasing
  - ➤ Operations
  - ➤ Legal
  - ➤ Human resources
  - ➤ Accounting and finance
- Information required by partners and investors
- Information required by contractors
- Design and technical information
- Cost information
- Project management information
- Surveys of information needs
- Use of information management models from past projects
- Use of information management models from software vendors

Information needs will be identified and documented as follows:

_____

_____

_____

_____

_____

_____

_____

_____

_____

_____

_____

_____

_____

| Refer to |
| Issue # |

C6: Communication Management Plan

e. What systems-based communication interfaces must be accommodated?

Things to consider:

- E-mail
- Electronic data interchange
- Global communications
- Owner and contractor intranet
- Telecommunications systems
- Project-specific communication systems
- Owner's communication systems
- Contractor's communication systems
- Need for "firewalls" and other forms of security

The project must accommodate the following communications interfaces:

_____

_____

_____

_____

_____

_____

_____

_____

_____

_____

_____

_____

_____

_____

_____

_____

_____

_____

| Refer to
Issue # |
| --- |

C6: Communication Management Plan

f. What are the issues with respect to Communication Management?

Enter each issue on the Issues List and Path Forward Planning Worksheet.

C6. Communication
Management Plan

VII. Issues List and Path Forward Plan

A successful Strategic Project Planning workshop will inevitably result in an extensive list of Issues and Action Items. During the workshop, the team fills in the first column of the Issues List and Path Forward Planning Worksheet, to ensure that no issues or action items will be lost.

Project teams always appreciate this critical last step in the workshop in which the issues and action items are consolidated and a Path Forward Plan developed to resolve them.

The Path Forward Plan is short-term, and usually focuses on the following 4 to 6 weeks.

Use the Issues List and Path Forward Planning Worksheet to develop the Path Forward Plan. Be sure to assign responsibility for leading each task. This worksheet can be updated from time to time as tasks are completed, issues resolved, and new issues identified.

Issues List and Path Forward Planning Worksheet					
Describe Issue or Action Item	Required Tasks to Resolve Issue or Complete Action	Person to Lead the Task	Status	Date to Complete	

Issues List and Path Forward Planning Worksheet

Describe Issue or Action Item	Required Tasks to Resolve Issue or Complete Action	Person to Lead the Task	Status	Date to Complete

Issues List and Path Forward Planning Worksheet				
Describe Issue or Action Item	Required Tasks to Resolve Issue or Complete Action	Person to Lead the Task	Status	Date to Complete

VIII. Using the Project Execution Plan as a Tool for Communication and Control

Many of today's top project managers have a hard time imagining how they would manage a project without a Project Execution Plan (PEP). The PEP:

- Ensures that everyone knows what the project's goals and objectives are

- Tells everyone what has to be done, when, by whom, and how

- Ensures that project management best practices are effectively applied

- Clarifies roles and responsibilities

- Helps identify and manage issues

The PEP is an extremely useful tool for *communication*. As the project progresses and new organizations and individuals come on board, studying the PEP is their first step in understanding what is important about the project. It is also the perfect way to communicate with management, since it has a strategic focus and relates to the business goals. In fact, many leading project organizations insist that a PEP be part of the materials reviewed by management prior to authorizing full project funding.

The PEP is a *living document*. Although we would love to have an ideal world in which everything held still for us until the project was done, in the real world nothing is constant. The project's business goals may change (e.g., as market conditions fluctuate), resulting in changes to project objectives, drivers, or scope. Milestones may therefore shift, organizations restructured, and contract strategies changed, and risk levels may rise or fall. If the PEP is kept current, then project plans will always be aligned with business conditions, and all participants aligned with the current goals and plans. Electronic publishing and distribution of the PEP is a good way to make it easy to update the PEP and distribute the updated information.

Since the PEP tells everyone what the project is about and the plans for executing it successfully, it is also an excellent tool for controlling project team performance. It sets objectives and defines a plan against which actual performance can be measured. We can use the PEP during project execution to assess:

- Are we likely to achieve our Vision of Success?

- Is our scope of work consistent with the plan?

- Are we meeting our major milestones?

- Are we meeting our cost objectives?

- Is our project team organization consistent with the plan?

- Is our contract strategy effective?

- Did we correctly identify risks, and is our risk mitigation plan successful?

- Are we managing human and material resources successfully?

- Are we executing the project in accordance with the PEP (that everyone had agreed to)? If so, how well is it working? If not, why not?

In a similar way, the PEP becomes a useful tool for conducting a post-project appraisal and capturing lessons learned.